REACHING OUT

Mobile STEM Modules
& Activities
for Students, Teachers, and Communities

ROBIN L. COOPER

Karen Mireau Books
Sonoma · California

ISBN: 978-1-968822-03-3

Cover Photos:
Ann-Simone Cooper O'Neil
Richard M. Cooper

Contents

Prologue

Over the past 30 years various activities in our lab group and colleagues have been involved in outreaching STEM activities with elementary-middle-high schools, public venues, life science days demonstrations, STEM camps, workshops for elementary and middle-high school teachers and CURE/ACURE courses at the college level.

Many of the modules presented herein were constructed for elementary school through college level with different activities in content but the same themes. The focus has been on keeping costs low and maintaining simplicity in the activities by using commonly obtained materials as well as using plants and invertebrates such as fruit flies for elementary to high school projects.

Also, an intentional goal was to make many activities mobile which can be taken to various classrooms and schools with minimal effort. We compiled modules easily to set up and be implemented in a STEM summer camp with and without access to electricity as well as modules which require a classroom setting. The grade levels are listed for each project or divided into sections within a module for the grade levels.

The chapters are truly a smorgasbord of various activities, but each one with a general theme. They are written with the help of various authors which have been involved in teaching or developing the content for the chapter. Publications related to the activities are listed as well as some content that the authors have put on YouTube.

Some of these activities have been published. Due to copyright, they are not fully reproduced in this book—instead, the content is briefly described so the reader can obtain the citation to the full descriptions.

— Robin L. Cooper
September, 2025

Chapter 1:

Examples of types of activities presented at various public venues and organized events

Many activities described below have been conducted at various venues. They have been useful in presenting STEM concepts, so they are just explained briefly and if there were related manuscripts they are cited.

These outreach activities were generally directed to promote healthy living. They have been presented in a flea market event, Lexmark STEM Day, Science Center in Louisville Kentucky, public schools life science days, and tabling at science fairs. When traveling overseas or on vacation maybe one can make it a working vacation to make connections.

Many of the events consisted of displaying human body organs of diseased states provided by the University of Kentucky Medical School for educational purposes from cadavers by donation to medical research. Human brains, spinal cords, lungs (i.e., from a smoker, coal miners), livers, hearts (fatty or enlarged), etc. were always a hit with curious middle- and high-school aged students.

Demonstrations were usually provided to illustrate concepts. Many of these demonstrations were also activities as listed in Chapter 2 depending on the audience. Some other activities are briefly described as follows:

1. $CO_2 + H_2O \leftrightarrow ... \leftrightarrow HCO_3^- + H^+$ and pH

Small containers (i.e., reused and cleaned *Drosophila* plastic bottles from college genetics classes or research labs) are half filled with distilled water and the other half with water mixed with instant ocean (2 cups for every 5 gallons; 473 ml for 18.9 liters). PH indicator Bromothymol Blue is used in each solution. Small amounts of 1M sodium hydroxide are added to the solutions until they turn dark blue.

Using a straw, participants exhale forcefully into the solutions. The distilled water solution quickly turns green (pH 7), and with continued blowing the solution turns light yellow (pH 5-6). The salt-water trial results in no color change. CO_2 reacts with water in the body to make acid. When one blows out (i.e. rids the body of CO_2), one rids the body of acid. When one retains the CO_2, the pH of the body decreases resulting in tissues not working correctly.

This is taught around the concept of the effects of smoking and Chronic Obstructive Pulmonary Disease (COPD). COPD is the most common lung disease, affecting over 12 million people in the United States.

For high school, college students, and teacher workshops, this is also a good time to teach about the misconception of pH only being in the range of 1 to 14 (due to only using a standard pH meter). For example, what is the pH of a strong acid of 10 molar? Also, there is the issue of temperature with pH and the full equation of pH (pH=- log[H+]a where 'a' is the activity coefficient.) As the temperature of water increases, the pH decreases.

2. International travel and use of BackYard Brain kits.

When traveling to other countries conducting research or collaborating with others, one might find a company sponsor to help purchase supplies for conducting educational outreach. It is helpful to have local contacts as the use of infrastructure from a company or using the locals to find a space can be accomplished. For example, Belize Aquaculture sponsored a high school teacher workshop while I was collaborating with them. The introduction of potential activities with kits from BackYard Brains was the focus of the workshop.

Many STEM concepts can be taught with the use of kits from BackYard Brains (https://backyardbrains.com/). Modifications of the initial intended use of the kits is also possible. Below is a series of activities a team of undergraduates have developed.

Various experiments and educational activities one can use with kits from BackYard Brains (see YouTube links).

Human muscle fatigue:
https://www.youtube.com/watch?v=e8BrVmYywZwand feature=youtu.be

Heart rate in a crab:
https://www.youtube.com/watch?v=oXgSkubpGdo

Sensory nerves in a leg of a crab/lobster/crayfish:
https://www.youtube.com/watch?v=yIT2rCvIUoo

Recording nerve activity with a suction electrode and Back Yard Brains. Part 1 of 2:
https://youtu.be/LrHXLF96d8Q

Use of suction electrodes to record from nerves in saline with Back Yard Brains devices. Part 2 of 2:
https://youtu.be/kGMBelEapGk

Measuring mechanosensory responses in the antenna with Spiker Box Back Yard Brains:
https://youtu.be/sQ0CGxkt3II

3. Activities with plants

Use of Backyard Brains for electrical recordings with cutting a leaf or turning on a plant light (with plant in dark box):
https://youtu.be/sEdBDbmVQ_s

4. Activities with invertebrates—crayfish HR, crayfish tail flip, *Drosophila* adult races, larval *Drosophila* crawling behavior (dark or light sides of a dish):

https://youtu.be/dnOfB-K3gyo
Recording electrical activity in plants

https://youtu.be/sGP_5ByY4NM
Crayfish heart rate while exposed to CO2

https://youtu.be/wTSynwmyHBQ
Making a simple and inexpensive microscope

https://youtu.be/wynMJjyTt1s
Mouth hook movements of larval *Drosophila*

https://youtu.be/smXe5axLZE8
Locomotion measures of larval *Drosophila*

https://youtu.be/zSJHN2NSrHk
Light and gravity sense in adults

5. Focus on being healthy—demonstrations with clear tubes and colored water

1st fluid flow: fundamental concepts—simulated "Ankle-Brachial index" (ABI):
http://youtu.be/XVr-MT3k0mw

2nd fluid flow: Laminar and turbulent flow:
http://youtu.be/KHxOwnh4YVo

3rd fluid flow: Viscosity:
http://youtu.be/ZOCNVUa0f_g

4th fluid flow: Windkessel effect:
http://youtu.be/UJt3-lGnhVU,
with student narration:

http://youtu.be/6iroS6arqT8, and
http://youtu.be/enZunzh7AnU

6. Types of events to participate at:

The flea market event; Lexmark STEM day; Science Center in Louisville for the public; public schools life science days; tabling at Science Fairs.

7. Structured activities in schools (details are presented in later chapters)

Optogenetics, *Drosophila* development, Physics concepts in biology, and Math with a biological relevance-stereology.

8. Promoting science fair projects over the years

- Growing plants and looking at the pattern of branches in stems and leaves in root related to compounds in the water.
- Plants growing in different soils
- Erosion control with different soils and types of ground cover
- No sleep affecting development/survival of adult *Drosophila* flies
- Stress on crayfish social behaviors and response to stimuli
- Insect behaviors
- Crayfish behaviors

9. CUREs in high school and college (details are presented in later chapters)

- Need for apprenticeship experiences for students to develop scientific practice skills, interest in science/STEM fields, research interest
- What is a CURE (whole class) -vs- Apprenticeship Research Experiences (ARE, research lab group activity).
- Environmental impact on behavior, development, physiology and survival of insects (*Drosophila*)
- Environmental impact on behavior, survival and physiology of crustaceans (crabs, crayfish)
- Themes: metabolic syndrome with fly development
- Population and food levels—ecology modeling.
- ACUREs in college level advanced physiology/ neurophysiology courses with lab infrastructure (details are presented in later chapters
- Research conducted in a lab group activity or course with students conducting authentic research

10.   Samples (photos at various activities):

11.1   The Healthy Flea Market.
Connected Science Learning November 2016-January 2017 (Volume 1, Issue 2)
https://www.nsta.org/connected-science-learning/connected-science-learning-november-2016-january-2017
By Robin Cooper, Kim Zeidler, Diane Johnson, and Jennifer Wilson

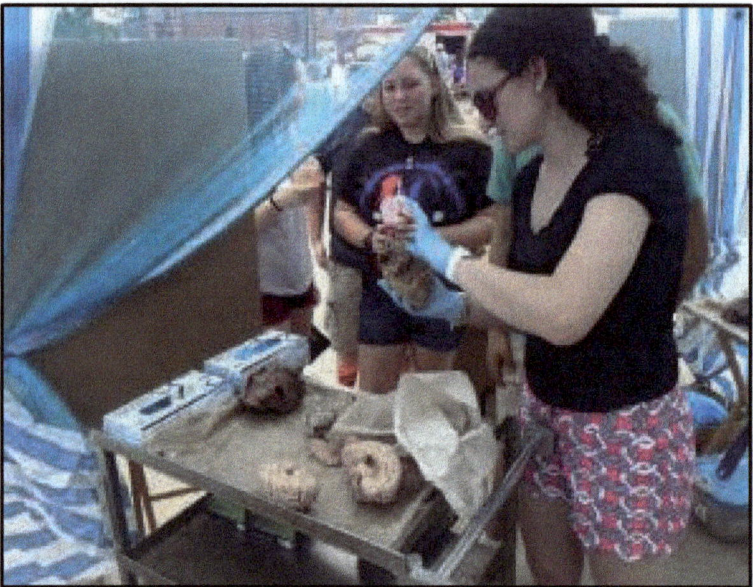

UK Project Engages Next Generation of Kentuckians to Tackle Health Problems.
http://uknow.uky.edu/campus-news/uk-project-engages-next-generation-kentuckians-tackle-health-problems.

11.2 Skype presentations with faculty members, postdocs, medical students with students in schools to talk about being healthy and the kind of research on going at the university (2015).

11.3 Hands-on workshop with high school science teachers in Belize with the use of kits from BackYard Brains. Left five kits with the teachers after the workshop. Sponsored by Belize Aquaculture.

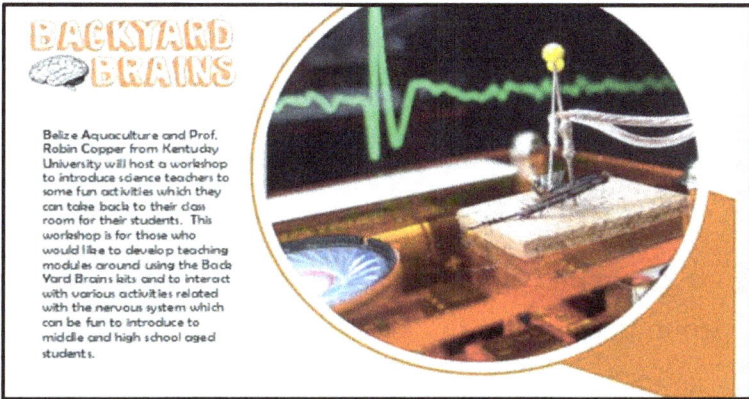

Chapter 2:

Presentations and activities
which can be conducted in a class period

Theme:
Sensory Systems

Level:
Elementary school level
and can be tuned to college level.

Short activities to be mixed and matched as time allows. Introduced to class by asking questions about how many senses humans have (5 is the common answer). One can joke and say the 6th sense is knowing what is going to happen in the future or reading their mind.

Smell & Taste

Ask about which animals have a good sense of smell. (look content up on Google)

To experiment on participants (Humans) use Hot Tamales or Red Hots candies.

Activity:
Have participants pinch nostrils closed (breath through mouth). Place 2 Hot Tamales in mouth and chew them up and swish them around in mouth. Then let the nostrils open and breathe in and out through the nose.

Results:

Younger participants (middle school) will generally sense this better than older participants. Some participants will state they "taste" it better after opening their nose.

Notes:

This is because about 80 % of what we think we taste is really what we smell.

This is why when one's sinus is stuffed up with a cold (flu), food generally does not have much taste. This is why good cooks use volatile spices to make their food taste better (i.e. smells better).

Citations:

Spence, C. Just how much of what we taste derives from the sense of smell? Flavour 4, 30 (2015). https://doi.org/10.1186/s13411-015-0040-2.

Hearing

Ask what animals can hear well.

An owl is a good example because one ear is lower than the other and feathers of different length on each side. This allows the owl to sense millisecond differences from right to left ear. Good for night hunting.

Activity:

Talk to an audience and ask them to cup their ears with palms facing the speaker, then one hand down, then the other hand. Now turn both palms away from the speaker and one hand down and then the other hand down.

Results:

Can hear much better with both hands up and facing the speaker.

Notes:

Why do humans not have bigger ears ? Which animals can turn their ears to face sounds ? (rabbits) What would an animal do if it could not turn its ears ? (turn its head instead). It was noted that the big ears of elephants have a lot to do with cooling the body (also noted for rabbits). So dual purposes of the ears. Also of interest is that it is now known that elephants communicate by stomping on the ground and elephants far away can feel the vibrations and respond.

Citation:

Stanford University. "Elephants Pick Up Good Vibrations -- Through Their Feet." ScienceDaily. ScienceDaily, 19 March 2001.
<www.sciencedaily.com/releases/2001/03/010312071729.htm>.

Sight

Activity:

Colors—Show color papers/objects (most people see colors red, blue and green, but some people can't because they are color blind.) Also, if one shows a red or a green square object on a classroom board from the projector for about 3 to 5 minutes, then switch slides to the same object but white, there is a delay and the color object still shows, as the sensory cells are still processing the previous color.

Determining the blind spot in the human retina.

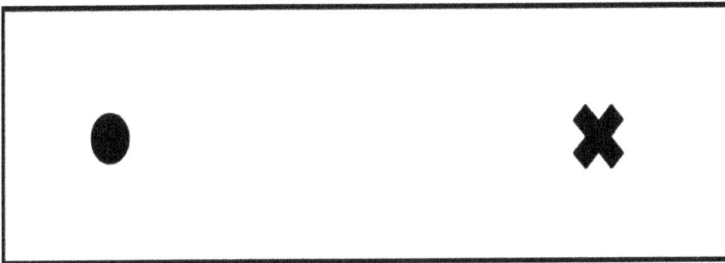

Close your left eye and look directly at the cross with your right eye. Move a card away from the face until the circle disappears.

Results:

The color lag. When one shifts the gaze to the white, the red-green receptors were still overstimulated, causing the green or pinkish afterimage.

The blind spot is due to the structure of the eyes where the optic nerve leaves the back of the retina. This is due to the anatomy of the retina where the nerves form in front of the sensory structures and then have to leave out of the retina. Interestingly, the eye of an octopus is arranged where the sensory structures face forward in the retina with the nerve forming out the back of the eye, so there is no blind spot.

Touch

A common activity is two-point discrimination from the fingertips to the upper arm.

Activity:
Use dull pointed toothpicks. Have the person being tested close their eyes and the 2nd person place two toothpicks at the same time at various distances apart ½ to 2 inches apart and see if the person states 2 or 1 toothpick. The touch on the skin has to happen at the same time and light touches.

Results:
On the fingertips one can readily tell two points apart due to small receptive fields (i.e., the area of skin a neuron monitors) and overlap of the receptor fields. The % overlap is less on the upper arm and larger receptive fields.

Notes:

For college level students one can jokingly say, now one can map out the whole body of their significant other and find out the most sensitive places for two-point discrimination. See . . . physiologists can have fun on Friday nights.

Let's start with least sensitivity to keep you in suspense . . . your back between your shoulder blades. Now for the highly sensitive regions . . .

Yeah, you guessed it . . . not something we can demo in class.

Sorry guys the females are much more sensitive due to smaller and more sensory fields on the clitoris then the penis.

Citations:

"The terminal innervation areas of individual neurons were 11 and 16 times smaller, respectively, in the clitoris than the penis, even though these neurons formed a similar number of corpuscles. This finding was in accordance with the 15-fold higher density of Krause corpuscles in the clitoris compared with the penis."

Fenner, A. Krause corpuscles act as genital vibration detectors.

Nat. Rev. Urol. 21, 455 (2024).
https://doi.org/10.1038/s41585-024-00922-7.

Now you have learned something new and practical.

Beyond the 5 senses
of the external world:

Proprioception

Activity:

A bodily illusion where the brain resolves a contradiction between the location of the nose and an outstretched hand, making the nose seem longer. This illusion can be created by having a participant touch the tip of their nose while receiving vibrations on their triceps under the back of the upper arm just past the elbow.

Have the participant start with eyes closed. Touch the index finger to the tip of nose, then slowly straighten arm out and back to nose. This occurs while the triceps are being stimulated with a muscle massage vibrator or even a vibrator from a science lab used to mix samples in a vial. Ask the person when they think they are about to touch their nose.

Results:

The person will state their finger is about to touch their nose when still 6 or so inches from their nose.

Temperature

It is odd that it is not listed as one of the external senses commonly thought of.

Activity:

Nail heads seem like they are easy to use and find for this activity. Roofing nails might seem like a good idea to use but the heads are too big. You need a flat head nail from what would normally be used to nail outdoor planks for fencing.

Have a metal coffee can filled with sand and placed in an oven to heat the sand. Place the head of the nail in sand with points sticking out. Put other nails in a tub of ice water. Test nail temperature on oneself before trying on a participant so as not to burn the participant.

Ask participants to close their eyes and state when they feel the hottest sensation as moving a nail head on the skin over a small area on the back of the hand. Use a red marker to place a dot where the hot sports are located. One has to keep trying new nails as they cool off. Repeat with a cold nail head after quickly drying the nail head off. Place blue dots where the cold spots are located.

Now take a warm nail head and place it on a blue dot and ask participants what they perceive.

Results:

The most sensitive places for heat detection appear along the blood vesicles in the skin. Sometime placing a hot object on a cold spot a participant will sense it as cold. This is called paradoxical cold.

Notes:

The cold sensory nerve goes to the part of the brain (CNS) that perceives cold, thereby stimulating it by warm or even an electrical stimulus it will be perceived as cold. Another example of hard wiring of neurons to perception is to go in a dark room for ten minutes. with eyes closed, gently pushing with one's finger on the side of the eyeball. Generally, one will perceive a crescent moon shape as the neurons where one is pushing are stimulated.

As a second activity mostly for college level students is to take a jalapeno pepper and cut it open so that one can rub the juice on the back of the hand. Why does it feel hot when the temperature is not different? The point is that the chemical in the pepper (i.e., capsaicin) stimulates the hot receptors and the hard-wired receptors relay the information to that part of the CNS for perception. When in the body capsaicin activates the set point neurons in the CNS (as same receptors as in the skin) and it tricks the body to think the body temperature is too hot, so the neural response is to try to cool the body off by sweating.

Chapter 3:

Understanding the human or animal body: Skeletal muscle and circulatory system

Theme:
Skeletal muscle

Level:
Elementary school level
and can be tuned into a college level activity

This was presented at the ABLE conference and published.

Cooper RL, Krall RM, Schultz MP, O'Neil AS, Dupont-Versteegden EE. (2020). Educational modules of skeletal muscle anatomy and function with models and active data gathering related to muscular dystrophy. Article 64 In: McMahon K, editor. Advances in biology laboratory education. Volume 41. Publication of the 41st Conference of the Association for Biology Laboratory Education (ABLE). https://doi.org/10.37590/able.v41.art64

In brief, the activities were as follows:

Anatomy

Various muscles from the grocery store and examine graphs:

1. If a compound microscope is available, take very small pieces (i.e., the size of a toothpick or even smaller) of uncooked chicken, pork, beef and fish.

Take piece on a glass slide. With another glass slide smash the meat down. Then take the slide off and place a cover glass slide on the tissue with the slide. Use clear fingernail polish to attach the cover glass to the slide. Now one can view with a compound microscope the banding pattern in the various types of meat. A 60X objective to an oil immersion 100X objective might be necessary.

Result:
A similar banding pattern can be observed.

2. Learn the anatomy of skeletal muscles.

   Begin with the history of the sarcomere and with cross sections of different bands. (2 D figures). List the vocabulary for a sarcomere so the students can label. How do the cross sections create a banding pattern? Have them create their models (2 D-3 D models). Possible use of PowerPoint for modeling.

   Use a generalized figure of skeletal muscle:

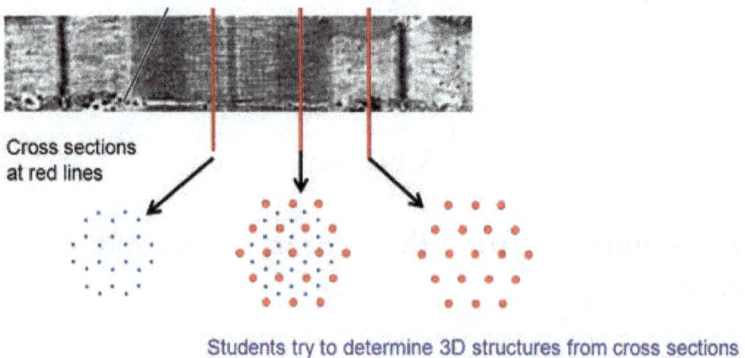

Cross sections at red lines

Students try to determine 3D structures from cross sections

Results:
There are many web based activities.

Participants at an elementary school to middle school level can build models with pipe cleaners of assorted colors.

Middle school students can add in anatomical parts for the whole muscle. (i.e. Titin, multiple units together in 3D.

Building a diorama of some aspects of skeletal muscle.

Activity:
Measure force of myosin and actin interaction (as a model).

Vernier force probes or a spring scale (fishing scale) will work for this activity. To demonstrate that the more Myosin head attached to actin, the stronger the connections are, and the more weight the unit can hold.

Activity:
Thematic approach with a storyline to understand skeletal muscle and a disease state.

This following subset of this module was presented as a workshop and is published by:

Cooper R. L., Krall R. M., Schultz M. P., O'Neil A. S., Dupont-Versteegden EE. 2020. Educational modules of skeletal muscle anatomy and function with models and active data gathering related to muscular dystrophy. Article 64 In: McMahon K., editor. Advances in biology laboratory education. Volume 41. Publication of the 41st Conference of the Association for Biology Laboratory Education (ABLE). https://doi.org/10.37590/able.v41.art64

Level:
***This content is designed
for High School instruction.***

***The foundation for this module
is the Middle School module.***

Too often life science is distilled into disparate facts addressing major biological concepts, but lacking purpose for learning and applying knowledge to real world contexts. The driving principle of this project is to build learning modules employing real world scenarios to foster authentic scientific investigation in biological science, in particular, the study of skeletal muscle cells and the effect of disease on cell function.

Authentic scenarios describing symptoms of a disease apparent in some adolescents anchors learning of cellular structure and functions and offer motivation for building knowledge on cellular functions. Content addressed in modules aligns with the Next Generation Science Standards (Achieve, Inc., 2013) for middle and secondary life science, Engineering Design, and Science and Engineering Practices, and promotes critical thinking skills in biological science.

Module activities begin with an introduction to the basic cellular structure and function, with a particular focus on skeletal muscle cell. Subsequent investigations include a review of the literature to learn the effect of disease on skeletal muscles, and creation of low- and high-tech 2D and 3D models to study effects of disease on healthy skeletal muscle cell functions. Vernier force probes and SketchUp (http://www.sketchup.com/learn), a 3D software program, are used to create high-tech models for studying and illustrating cell dysfunctions inherent to the targeted disease.

The module presented in this presentation uses Muscular Dystrophy as the anchor for exploring its effects on skeletal muscle cell functions, and subsequent observed effects. Muscular Dystrophy was selected as a targeted disease because of its prevalence in adolescents in middle, secondary, and college levels. Many healthy adolescents may know peers with the disease but lack understanding of its effects on the body. Further studying skeletal muscle cells in this context requires in depth study of the whole cell and its relationship with the plasma membrane and whole tissue to explain the disease.

Modules are designed in a multilevel approach for easy adapting to middle, secondary, and college levels. Pre- and post-testing with treatment and control groups will be used to evaluate changes in students' content knowledge of cellular functions, effects of disease on these functions, and science and engineering practices.

Part 1 is a class problem to work through.

Part 2 is to go through the middle school modules 1-3 and to compile an overview of the different forms of Muscular Dystrophy.

Part 3 is to build a physical model to help understand the problem with Duchenne's Muscular Dystrophy. The netting around the sarcomere drawing can be pulled and relaxed to feel the differences.

Part 4 is to use the free version of SketchUp (http://www.sketchup.com/learn), and model in 3D some aspects of skeletal muscle anatomy.

Part 5 is to talk or communicate with physicians and/or researchers about skeletal muscle disorders:

Ask about their research (basic or applied).

Types of muscle diseases the person might have seen recently.

What do they foresee in the future for research related to skeletal muscle ?

Part 6 is to investigate the comparative differences in skeletal muscle from different animals. Compile a report on the differences with diagrams and/or tables from the literature research.

Chapter 4:

Understanding the human
or animal body

Theme:
Circulatory system

Level:
High school to college

This was presented as a poster previously at an ABLE conference. In brief, an overview and supplementary material is presented in the manuscript below.

Stanley C, Krall R. M., Zeidler-Watters K., Johnson D., Blackwell R. R., Cooper R. L. (2020). STEM and health: stressors on the circulatory system. Article 82 In: McMahon K., editor. Advances in biology laboratory education. Volume 41. Publication of the 41st Conference of the Association for Biology Laboratory Education (ABLE). https://doi.org/10.37590/able.v41.art82. https://www.ableweb.org/biologylabs/wp-content/uploads/volumes/vol-41/82_Stanley.pdf.

The goal of this problem-based module is to explore the effects of health-related issues (e.g., obesity, arteriosclerosis) on pressures in the circulatory system. Modeling and engineering design are key practices comprising the Next Generation Science Standards, making these modules timely and well-suited for life science classrooms at the introductory college level. These STEM modules bridge standard-based biology, physics, and health concepts in an integrative approach to learn fluid dynamics and physiology in authentic situations.

Physiological issues represented include: (1) plaque formation and effect on flow, (2) elastic recoil and arteriosclerosis, (3) effects of blood viscosity on flow, and (4) differential blood pressure related to resistance.

Exercises guide learners in assembling and using human circulatory system models to explore principles of pressures as they relate to tubing, flow, and resistance. Findings from the investigations are used to construct diagnoses and recommended treatments for patients in realistic scenarios with a secondary emphasis on a healthy lifestyle.

This module in the past was also addressed as a conceptual module for a classroom setting which went over well with the students. This is shown below.

This was developed by Dr. R. M. Krall (Department of STEM, University of Kentucky).

Cindy, a 40-year-old high school teacher, is planning a summer trip with her teenage nieces trekking along the

Continental Divide Trail in Colorado and Wyoming. According to the Trails.com website http://www.trails.com/continental-divide-trail.aspx:

> *The Continental Divide Trail might well be the most extreme of the three major National Scenic Trails. The "CDT" covers the greatest distance at approximately 3,100 miles, reaches an ultimate high point of 14,270 feet at Grays Peak, Colorado, and has a low point of 4,280 feet, less than 2,400 feet below the highest point on the Appalachian Trail.*

The trip Cindy and her nieces have selected will be very strenuous and in high altitude. Is Cindy healthy enough to safely complete the trip? That is the question her physician must answer.

Cindy's Health History

In high school, Cindy ran cross-country track and was in very good shape. After college she became a high school teacher in Lexington, Kentucky (elevation 978 feet) and has not had much time to exercise and stay in good physical health. She has not been eating the most nutritional meals over the last several years and as a result has gained considerable weight. Cindy really wants to go on this trip with her nieces, but she wants to make sure she is "healthy enough" to do so. She has scheduled a visit with her physician for a checkup before beginning her training for the trip.

Cindy's Physical Check Up

Upon learning of Cindy's plans to hike the Continental Divide, the physician performed a complete physical. She measured Cindy's weight and blood pressure, checked her pulse, listened to her lungs as Cindy breathed, tested her reflex reactions, and examined her abdominal region. The physician drew blood and ordered a complete blood workup, including a complete lipid panel to check Cindy's cholesterol and triglycerides. The physician was concerned because Cindy was 5'5" and weighed 175 pounds, bordering on obesity levels. She was greatly concerned about Cindy's ability to function normally in the high altitudes of the Continental Divide.

The results from the physical were of concern. A comparison of Cindy's blood pressure at her ankle and brachial arteries measured greater than 15mmHg difference (ABI, ankle-brachial index, measured by dividing the higher of the two blood pressure measurements at the arms by the higher of the two measures near the ankle). Cindy' blood pressure in the right arm is 160/95 and she is considered overweight by 40 pounds for her height and body structure. The physician ordered a Doppler ultrasound and a stress test to assess her circulatory and cardiac function.

Upon her follow-up visit, the physician reported to Cindy the lab and test results along with specific recommend-ations. Cindy's blood analysis revealed a HDL of 40 mg/dL and an LDL of 165 mg/dL. The total cholesterol is 210 mg/dL. The hemoglobin level was 11 gm/dL with a hematocrit of 36%. The ABI was 0.80, which indicated poor blood flow to the ankle or brachial arteries. Doppler ultrasonography revealed poor circulation in Cindy's femoral and carotid arteries. Cindy's right femoral artery was about 50% occluded and her carotid arteries were both about 25% occluded due to plaque buildup. The physician suggested a conventional angiography of her circulatory system to examine for severe problem points in her circulatory system as well as examine her coronary arteries because she performed poorly on the treadmill stress test.

The physician advises Cindy that she is not physically fit in health for the strenuous hiking trip along the Continental Divide at such an altitude. The physician recommends medications to start reducing the plaque buildup in her arteries and dietary control of lipid and cholesterol intake. The physician also suggests another test to determine if Cindy has a nutritional deficiency in iron and B12 or if she has an underlying disease responsible for the low hemoglobin and hematocrit values.

Designing a Demonstration Model

The group of students had to design the most cost efficient and effective model to demonstrate the main points related to health conditions.

YouTube links:

1st fluid flow:
fundamental concepts—simulated "Ankle-Brachial index" (ABI)
http://youtu.be/XVr-MT3k0mw

2nd fluid flow:
Laminar and turbulent flow
http://youtu.be/KHxOwnh4YVo

3rd fluid flow:
Viscosity
http://youtu.be/ZOCNVUa0f_g

4th fluid flow:
Windkessel effect
http://youtu.be/UJt3-lGnhVU
with student narration

http://youtu.be/6iroS6arqT8 and
http://youtu.be/enZunzh7AnU

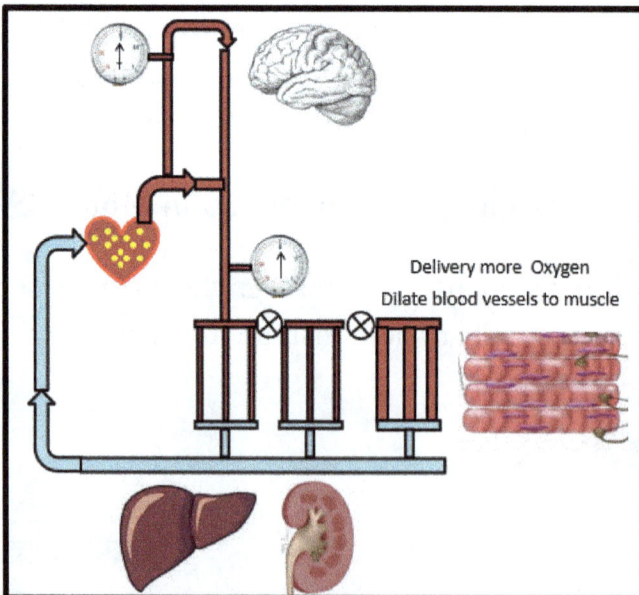

Delivery more Oxygen
Dilate blood vessels to muscle

Chapter 5:

Revisiting Mendel:
Use of a behavioral assay
to examine inheritance of traits
in *Drosophila*

Level:
High school and college

Most of this content was published. A summary is below. This activity helps to allow students to conduct their own genetic crosses and learn to determine sex of adult *Drosophila*. The use of optogenetics presented an opportunity for the students to investigate and understand the power of optogenetics to help humans with various disease states as well as the basics of how gene theory could potentially be used.

Jeffrey M. Chalfant, Robin L. Cooper, Tawny Aguayo-Williams, Lexie Holtzclaw, Madison Loveless, Jennifer Wilson, and Doug Harrison. 2022. Revisiting Mendel: Use of a Behavioral Assay to Examine Inheritance of Traits in *Drosophila*. Article 56 In: Boone E. and Thuecks S., eds. Advances in biology laboratory education. Volume 42. Publication of the 42nd Conference of the Association for Biology Laboratory Education (ABLE).

https://doi.org/10.37590/able.v42.art56

Using the established rules of Mendel and others, predicting the outcome of genetic crosses in model organisms is a common exercise for college students. Frequently, one uses visible phenotypic markers such as curly wings, eye color, and abnormal bristles to identify genetic outcomes. Yet many genetically based traits, such as behavioral and physiological characteristics, are not easily observed. To demonstrate that such traits can likewise display classical genetic inheritance, we utilized an optogenetic system in *Drosophila* to modify response to light.

We utilized the inheritance of behavioral responses associated with light-activated channel rhodopsin in motor neurons and body wall muscles. The frequency of responsive animals was quantified over multiple generations beginning with two pure-breeding (homozygous) strains, each containing one of the two components needed to produce the light-sensitive proteins.

The use of light-sensitive channels to examine the predicted genetic outcomes is an approach which can be used in teaching classical genetic principles using non-traditional phenotypes. Green fluorescent protein (GFP) can be expressed to illustrate which cells are expressing channel rhodopsin. This introduces concepts of transgenesis, genetically modified organisms, and genetic contributions to behavior. In addition to basic dominant and recessive allelic relationships, the experiments introduce more complex genetic concepts, such as epistasis, gene expression and cellular diversity, as well as physiological and behavioral traits of animals.

This module is presented in a variety of ways depending on equipment availability and can be used in a hybrid or remote format with data provided.

The students were able to confirm the inheritance of a single GFP transgene.

Examine the inheritance of ontogenetically-regulated behavior controlled by two transgenes.

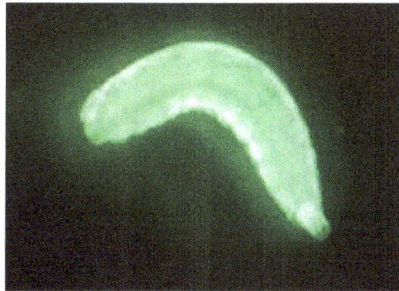

Expression of GFP can also be observed in pupae instead of larvae. The muscle expression with the MHC-GFP is seen below.

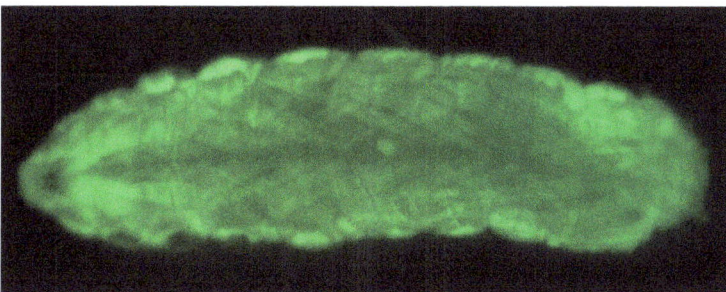

The module also explains how to build one's own LED light source.

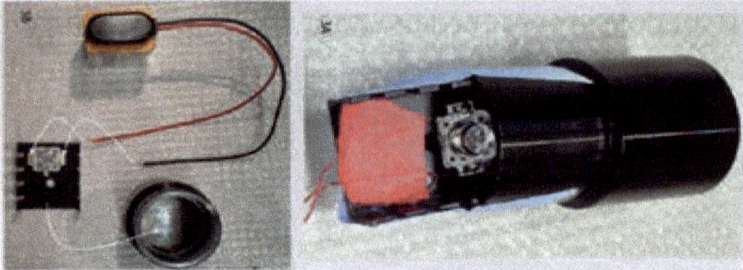

A schematic illustration of the expected results of two generation intercross of a GAL4 driver (GAL-24B shown) with the UAS-ChR XXL optogenetic responder is shown below. The transgene inheritance is dominant for each, but a larva must have at least one copy of each transgene to display the paralysis behavior. This epistatic relationship should give rise to 9/16 of progeny in the F2 generation that show the behavioral trait.

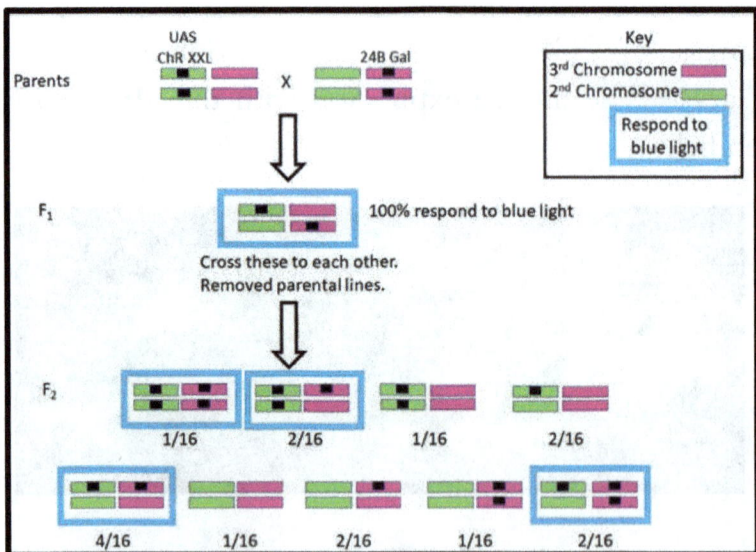

Key Citations for this module:

Bradforth S. E., Miller E. R., Dichtel W. R., et al.: University learning: Improve undergraduate science education. Nature. 2015, 523(7560): 283–284.

Dawydow A., Gueta R., Ljaschenko D., Ullrich S., Hermann M., Ehmann N. et al. Channelrhodopsin-2-XXL, a powerful optogenetic tool for low-light applications. Proc Natl Acad Sci. 2014, 111(38): 13972–13977.

Majeed Z., Koch F., Morgan J. *et al.* A novel educational module to teach neural circuits for college and high school students: NGSS-neurons, genetics, and selective stimulations. F1000Research 2017, 6:117. https://doi.org/10.12688/f1000research.10632.1

Pulver SR, Hornstein N. J., Land B. L., et al.: Advances in Physiological Education. 2011; 35: 82–91.

Titlow J.S., Anderson H.., Cooper R. L. (2014). Lights and Larvae: Using optogenetics to teach recombinant DNA and neurobiology. The Science Teacher National Science Teacher Association, 81, 3–9.

Chapter 6:

Bridging optogenetics, metabolism, and animal behavior for student-driven inquiry

Level:
High school and college

This module was published and is explained in brief below. The student learning objectives were:

Students will be able to explain what a transgenic organism is.

Students will be able to form hypotheses on how optogenetics can be used to manipulate locomotion in *Drosophila melanogaster* larvae.

Students will be able to explain how temperature and cofactors play roles in metabolic processes.

Students will be able to describe practical applications of biotechnology, particularly opto-genetics.

Tawny Aguayo-Williams, Vaaragie Subramaniam, Doug Harrison, Robin L. Cooper, and Brett Criswell. 2022. Bridging optogenetics, metabolism, and animal behavior for student-driven inquiry. Article 52 In: Boone E. and Thuecks S., eds. Advances in biology laboratory education. Volume 42. Publication of the 42nd Conference of the

Association for Biology Laboratory Education (ABLE).
https://doi.org/10.37590/able.v42.art52.

Overview

This module is primarily a teaching tool for advanced undergraduate students in physiology and genetic courses that participate in experimentation. The significance of this exercise is that it can lead to deeper levels of discussion in the mechanism behind how channel rhodopsin functions as ion channels (i.e. properties of ion channels) as well as how temperature influences metabolism and gene regulation can all be very complex and difficult to address without some reading of primary literature. One may wish to leave it in terms of generality depending on the level of the course the participants are enrolled in.

Biotechnology is an ever-evolving field of science critical to improving the quality of human life, particularly in medicine. Optogenetics, an area of biotechnology, involves genetic modification of cells to express light-sensitive ion channels, which allows for the use of light to manipulate behavior. This module utilizes an approach to bridge optogenetics, cellular metabolism, and animal behavior for student-driven inquiry in college courses. *Drosophila melanogaster* larvae modified to express Channelrhodopsin-2 (ChR2) in motor neurons serve as model organisms in this module. Students can connect temperature, metabolic rate, and gene expression through data collection of behavioral responses to light stimuli exhibited by larvae raised at various temperatures. Students can observe the role of cofactors in metabolic processes via larvae that have been fed all-trans retinal, a cofactor to ChR2. Students can analyze and interpret data in order to make a claim about how the two variables investigated (temperature and cofactor) impact the behavior of the target organism. The final activity allows students to form connections to cutting-edge research related to optogenetics, such as work with the GAL4/UAS system. This would provide a foundation for further exploration of such research. The module can be used in in-person, hybrid, or remote settings.

Sample of background information provided in the module:

"The behavior of animals is controlled by complex physiological processes involving all the bodily systems. A novel approach is in the expression of light activated proteins in defined tissue or cells to excite or depress the excitability of defined cells to control a tissue such as a heart or skeletal muscle and even defined neural pathways with the central and peripheral nervous systems (Gunaydin et al., 2010; Deisseroth, 2015). The fruit fly, *Drosophila melanogaster*, is especially attractive for such studies because of the easy manipulation of cell specific gene expression and the reproducible stereotyped behaviors exhibited. Both larval and adult forms of *Drosophila* have been utilized as a proof of concept in activation of light sensitive proteins. Since genetic crosses are used for expression and light is used to activate the light sensitive proteins, the term of optogenetics is used to describe this technique (Deisseroth, 2015; Han and Boyden, 2007; Fenno et al., 2011; Camporeze et al., 2018; Boyden et al., 2005). The crawling of a larvae or reduced climbing on a tube for adults can be readily quantified for alterations in response to altered neural control of muscle or altered activity of the body muscles themselves. The during and intensity of these behavioral alterations can also be readily quantified by direct visual observation."

As part of the graphical and data literacy portion of this module, we recommend having students create a graph of the data that they have collected that can be shared during the discussion portion. We suggest giving them the freedom to be as creative as they would like with this process. Any errors within their graph can be addressed during discussion. This is also a great opportunity to incorporate technology for students to utilize to create their graphs. Some suggestions include Graphical, Infogram, and Plotly.

Key citations:

Boyden E. S., Zhang F., Bamberg E., Nagel G., Deisseroth K. Millisecond-timescale, genetically targeted optical control of neural activity. Nat Neurosci. 2005;8(9): 1263–1268. pmid:16116447.

Camporeze, B.; Manica, B. A.; Bonafé, G.A.; Ferreira, J. J. C.; Diniz, A. L.; de Oliveira, C. T. P.; Mathias, L. R., Jr.; de Aguiar, P. H. P.; Ortega, M. M. Optogenetics: The new molecular approach to control functions of neural cells in epilepsy, depression and tumors of the central nervous system. Am. J. Cancer Res. 2018, 8:1900–1918.

Dawydow, A.; Gueta, R.; Ljaschenko, D.; Ullrich, S.; Hermann, M.; Ehmann, N.; Gao, S.; Fiala, A.; Langenhan, T.; Nagel, G.; et al. Channelrhodopsin-2-XXL, a powerful optogenetic tool for low-light applications. Proc. Natl. Acad. Sci. USA 2014, 111:13972–13977.

Deisseroth, K. Optogenetics: 10 years of microbial opsins in neuroscience. Nat. Neurosci. 2015, 18:1213–1225.

Han, X.; Boyden, E. S. Multiple-color optical activation, silencing, and desynchronization of neural activity, with single-spike temporal resolution. PLoS ONE 2007, 2:e299.

Higgins, J., Hermanns, C., Malloy, C. and Cooper, R. L. (2017) Considerations in repetitive activation of light sensitive ion channels for long term studies: Channel rhodopsin in the *Drosophila* model. Neuroscience Research 125:1-10.

Titlow J. S., Anderson H., and Cooper R. L. (2014). Lights and Larvae: Using optogenetics to teach recombinant DNA and neurobiology. The Science Teacher National Science Teacher Association, 81, 3–9.

Chapter 7:

A hands-on educational module to teach aspects of human dietary health using fruit flies as a model.

Level:
High school and college

Aspects of this module were presented at a meeting virtually due to COVID only as an oral presentation.

Brittany L. Slabach and Robin L. Cooper. 2022. An active learning approach to teach aspects of human dietary health using fruit flies as a model. Article 39 In: Boone E. and Thuecks S., eds. Advances in biology laboratory education. Volume 42. Publication of the 42nd Conference of the Association for Biology Laboratory Education (ABLE).
https://doi.org/10.37590/able.v42.art39
https://www.ableweb.org/biologylabs/wp-content/uploads/volumes/vol-42/39_Slabach.pdf.

Instructor Notes:

This module highlights the relationship between diet, development and behavior using *Drosophila melanogaster* (fruit flies) as a model. Fruit flies are a commonly used model organism in research to understand biological principles and became well recognized as a model for studying genetics (Rubin and Lewis 2000; Morgan 1910). Using a project-based learning approach and building on the tremendous amount of knowledge about the life cycle, physiology, genetics and behavior of *Drosophila,* this module provides students the opportunity to investigate questions concerning chronic health of their interest.

The module was developed to allow for flexibility in content investigated by building on the same two foundational modules: (1) the influence of diet on development and (2) the effects of diet on behavior. Therefore, the module can be implemented as consecutive units, as it's presented here, or as stand-alone single units. Additional concepts can be explored using these two modules as building blocks. These concepts include survival of the adults and population dynamics (Oh and Oh 2011, Potter et al. 2016, Pulver et al. 2011); the effects of a ketogenic diet on behavior and function related to treatment of epilepsy (Boison 2017); and using heart rate as a bioassay for health of the larvae exposed to various diets (Spindler 2005, Potter et al. 2019).

Each aspect of this module is highlighted in a series of student made movies listed below:

Individual educational units for the module

Movies provided go along with introductory text, protocol text and sample graphs of data, as well as notes to instructors for each activity. Movie order:

1. Intro to "A model for learning about aspects of metabolic syndrome"
(LaShay Byrd)
https://youtu.be/FZ1kB1_9QMM.

2. Diets for metabolic syndrome
(LaShay Byrd and Jenni Ho)
https://youtu.be/22Onri7mPdg.

3. Stages of larvae (Jenni Ho)
https://youtu.be/ZVMnsA2o44U
Collecting the staged larvae
https://youtu.be/vJ7ZV0hxM5g.

4. Making a simple microscope
(Ruth Sifuma)
https://youtu.be/wTSynwmyHBQ.

5. Body wall movements
(Brecken Overly)
https://youtu.be/smXe5axLZE8.

6. Mouth hook movements
(Hunter Maxwell and Crysta Meekins)
https://youtu.be/wynMJjyTt1s

7. Ethograms and why ethograms
(Brittany Slabach)
https://youtu.be/8MMOJj4nMY0.

8. Hat Assay
(Maddie Stanback and Emma Rotkis)
https://youtu.be/fG7iFRF9HDg.

9. Development of fruit flies-pupation rate
(Clare Cole and Kay Johnson)
https://youtu.be/qoFhLFie3K0.

10. Anesthetize adult flies for moving them
(Samantha Danyi)
https://youtu.be/ZfbN1GTu-Gg.

11. Light and gravity sense in adults
(Sushovan Dixit)
https://youtu.be/zSJHN2NSrHk.

12. Effects of diet-ketogenic specific
(Madan Subheeswar)
https://youtu.be/UiPDIEDa_mk.

13. Measuring heart rate in larvae (Ann Cooper et al., 2009)
Cooper, A. S., Rymond, K. E., Ward, M. A., Bocook, E. L. and Cooper, R. L. (2009) Monitoring heart function in larval *Drosophila melanogaster* for physiological studies. Journal of Visualized Experiments (JoVE) 32.
http://www.jove.com/video/1596/monitoring-heart-function-larval-drosophila-melanogaster-for.

14. The effect of thermal stressors on larvae and adults fed various diets. (Alexandra Stanback)
https://youtu.be/lnZlK_YniqM.

15. A summary of this educational module
(LaShay Byrd)
https://youtu.be/dBKUSNyjub8.

16. Collecting eggs and staged larvae
https://youtu.be/vJ7ZV0hxM5g.

17. A simple way to knock out flies to transfer a given number to another culture vial.
https://youtu.be/sGP_5ByY4NM.

Materials & Supplies:
Our goal is for college level programs of all budget types to be able to use this model. We have provided information on all necessary materials and supplies including where materials can be purchased and alternative options in the Appendix.

Food Preparation and Fly Diets:
To maintain adult flies for breeding and rearing, a cornmeal-molasses-agar media can be made. A complete ingredient list and instructions can be found for the media in the Appendix. Various additions can be included to this standard media to investigate the effects of different diet types. For example, if one wants to examine essential amino acids in a diets various amounts of amino acids could be used and in combinations. Different *Drosophila* lines with mutations are available to investigate defects in amino acid transport and enzymes used in metabolism (St Clair et al., 2017; Sasamura et al, 2013).

To focus on diets related to human health, one or more of three different diets can be used for this activity: (1) high fructose, (2) high protein (soybean extract), and (3) high fat (coconut oil). These additions can be included in the standard media in 5%, 10%, 20%, 40% per weight of diet. A higher fat diet than 40% is difficult to use as the fat does not mix well with the food. Standard fly media food can be used and mixed to these percentages based on wet weight of the readymade food. This is the easiest approach in our experience.

Fly Growth:

The two activities presented here are set up sequentially, therefore setting up a duplicate vial will allow one to fully pupate and the other to be tested for behavioral differences. Eggs, 1st instar, or 2nd instar larvae can be used to set up the experiments. Growth from egg to third instar larva takes approximately five days (Figure 1), and larval stage can be determined by anatomical and behavioral differences. Adult flies can be exposed to each food type allowing them to lay eggs, then eggs can be moved to hatch on different food types. This would require a minimum of a four-hour time window.

In this approach it is difficult to know the number of viable larvae one is stating out with. However, one could use control food and compare relative to the controls. For more precise measures of survival (experiment one), we strongly suggest moving 1st or 2nd instars into various types of food. It is easy to damage the larvae, so care is required when transferring them. We suggest having a fly colony started on traditional media to make it easier to set

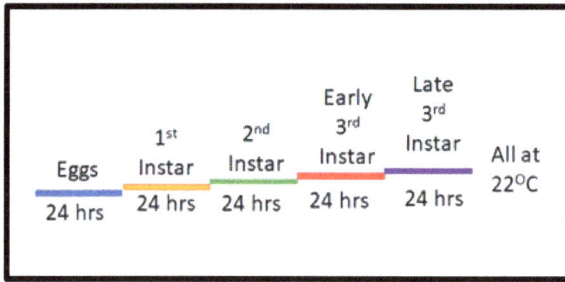

Figure 1.
**Developmental timeline of *Drosophila*
at 22 degrees Celsius.**

up the activity and provide flexibility in timing. A procedural outline for establishing a fly colony and preparing individual vials for the activities can be found outlined in the Appendix.

Additional Student Engagement

For additional engagement regarding the context of the activity, a pre-lab assignment where students investigate the health status of their local community could be implemented. Using a CDC website, students can examine the rates of cardiovascular disease for example. This can spark interest in the topic and help provide more specific background for students to help frame the concepts for students.

Student Handout

Learning Objectives:

- To explore the role of diet in physiological and developmental processes

- Generate informed hypotheses regarding the effects of diet on the development and behavior of *Drosophila melanogaster*

- Design and conduct an experiment to test the hypotheses

Background Information

Diet and Health:

The general notion is the balance of "energy in" with the "energy out" for the developmental and adult life requirements of organisms for being in a healthy state. Excess "energy in" can led to storage of the energy in the form of fat or surplus circulating levels of substances which can have harmful consequences, such has high lipids and sugar content in the cardiovascular circulation. Therefore, metabolism and diet type are keys factors in the energy homeostatic balance. And we use different diet types for both medical treatments for purely cosmetic reasons. For example, body builders commonly use a high protein – low carbohydrate diet to build and maintain lean muscle mass. Yet diet can also be used to control our cholesterol or blood sugar levels, and a ketogenic diet (high-fat, adequate-protein, low-carbohydrate diet) is also being used to treat epilepsy. We can investigate the effects of diet on physiological processes, mimicking the different factors observed in human conditions, but altering the diets of *Drosophila.*

Life cycle of *Drosophila*:

After males fertilize the eggs and the female lays them on a substrate in 1 day the larvae will emerge at room temperature (21°C or 70°F). Larvae develop through three stages which are easily identified by morphological changes in their mouth hooks used for feeding (Figure 1). In the late 3rd instar stage the larvae crawl out of the food and find a place to become a pupa. After about 7 days the pupa emerges as adult fruit flies. The adults typically live for 2 to 3 weeks depending on the crowding and environmental conditions.

Figure 1.
Development life stages of *Drosophila*.
(See a figure from Van Timmeren et al., (2017).

Diet Options:

1) High Fructose—A condition which is of increasing prevalence in the USA and other industrial nations is that of metabolic syndrome. This condition results in an increased blood pressure, high blood sugar, excess body fat around the waist, and abnormal cholesterol or triglyceride levels, increasing the risk for heart disease, diabetes, and stroke. To mimic some of these factors in *Drosophila* we can use a high sugar diet with alpha-fructose, a simple sugar.

2) *High Fat (Ketogenic)*—A diet high in fat can also result in metabolic syndrome. Therefore, to mimic these factors in *Drosophila* we can use a high fat diet by using various amounts of 100% coconut oil.

3) *High Protein*—To mimic a diet high in protein, like what a body builder may use, we can use synthetic soybean extract as a protein source.

Choose a diet from the list above
and develop a hypothesis regarding how that diet may affect
the development of Drosophila. State your hypothesis.

How might the diet influence the behavior of the larva?
State your hypothesis.

Now let's set up the experiment to test your hypotheses.

Experiment One

Influence of Diet
on the development of *Drosophila*

To test the effects of diet on larval development, we will compare larvae grown on a high fat and high sugar diet. Here we will set up replicates of each tube. We can use one of the replicates to investigate the effects of diet on behavior (Experiment Two).

Materials:

    2 tubes of 10% high fat larval diet
    2 tubes of 10% high sugar larval diet
    2 tubes of standard fly media
    1 dish of 2nd or 3rd instar larvae
    Paint brush/forceps (to move flies)
    Permanent marker

Procedure:

1. Place 10 larvae (1st or 2nd instar larvae) into each of the high fat and high sugar larval tubes

2. Be sure to note what instar they are and how many of each you include in the tube

3. Allow larvae to burrow into food ($\sim$ 30 minutes)

4. Mark and index the location of all larvae using a sharpie as shown in Figure 2 (right). Note date and time pupa form in notebook.

5. Place tubes in a light-controlled environment at 22 degrees Celsius for 24 hours

6. Check every 24 hours to determine the number of individuals that pupate

7. Note any changes in pupation and survival in your notebook.

What is our dependent and independent variable?
What is our control?

What is the percent survival based on diet type?
Do you accept or reject your hypothesis?

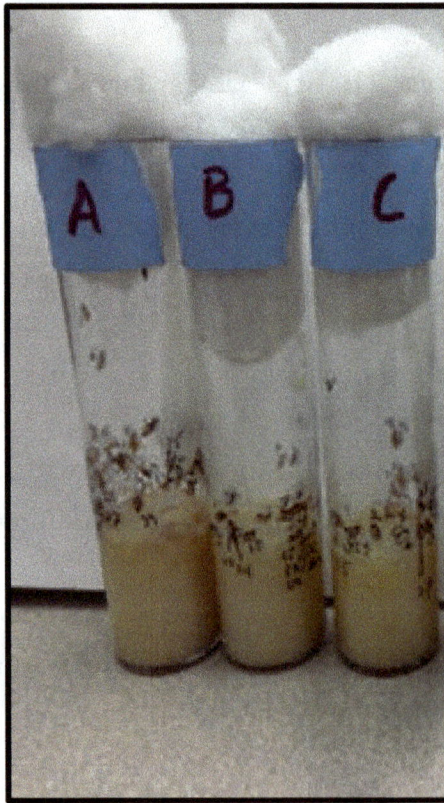

Figure 2:
**Marked and indexed larvae in different diet types.
Diets were labeled using a lettering system,
and all larvae were circled and marked
using a permanent marker.**

Experiment Two
Effects of Diet on Behavior

Next let's investigate the effects of diet on behavior. But first we need to understand different behaviors of *Drosophila* larvae. Larvae respond to external stimuli in a variety of ways. These behavioral responses are adaptive and act as an anti-predator defense (Robertson et al. 2013). Prior to proceeding, review the different behaviors that larvae can exhibit (see Appendix).

Develop a hypothesis
to explain how each diet might influence
larval behavior

Materials:

From Experiment One:

    1 tube of 10% high fat larval diet from experiment one

    1 tube of 10% high sugar larval diet

    1 tube of standard fly media

    Tweezers or forceps to move larva

    Petri dish

    Pencil

    Stopwatch

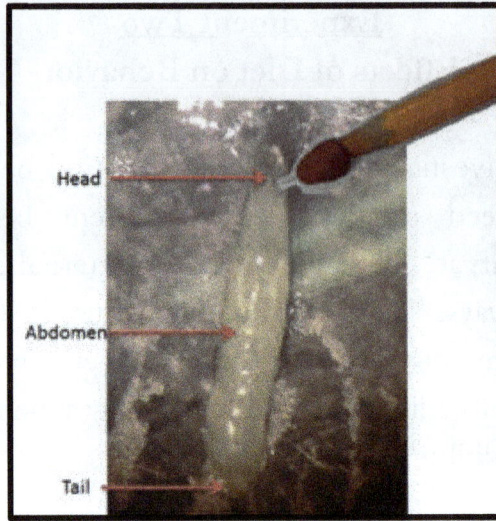

Figure 3.
**Illustration of body segments on larva
and points of contact.**

Procedure:

1) Using tweezers or a small paint brush remove a single larva and place it on a petri dish

2) Identify the stage of the larva

3) Allow larvae to acclimate for 15 secs prior to beginning the trial

4) Count the number of peristaltic waves for 15 seconds

5) Stimulate the larvae at the abdomen on the right side on the larva's body segment as shown in Figure 3

6) Record the behavioral response

7) Wait 15 seconds

8) Repeat steps 5-6 for each additional body segment

9) Conduct stimulus test on a minimum of 10 larvae per diet type and control

When all stimulus tests are finished, calculate the fraction of each behavioral response by diet type.

Do you accept or reject your hypothesis?

Citations:

Boison, D. (2017). New insights into the mechanisms of the ketogenic diet. Current Opinions in Neurology. 30:187-192.

Morgan, T. H. Chromosomes and heredity. The American Naturalist. 44 (1910):449–96. http://www.jstor.org/stable/pdf/2455783.pdf. (Accessed March 25, 2017).

Oh, P. S. and Oh, S. J. (2011). What teachers of science need to know about models: an overview. International Journal of Science Education. 33:1109–1130.

Potter, S., Krall, R. M., Mayo, S. Johnson, D., Zeidler-Watters, K., and Cooper, R. L. (2016). Population dynamics based on resource availability and founding effects: Live and computational models. The American Biology Teacher. 78(5): 396-403, ISSN 0002-7685.

Potter, S., Sifers, J., Yocom, E., Blümich, S. L. E., Potter, R., Nadolski, J., Harrison, D.A. and Cooper, R. L. (2019) Effects of inhibiting mTOR with rapamycin on behavior, development, neuromuscular physiology and cardiac function in larval *Drosophila*. Biol. Open 2019, 8.

Pulver, S. R., Cognigni, P., Denholm, B., Fabre, C., Gu, W. X. W., Linneweber, G. et al. (2011). Why flies? Inexpensive public engagement exercises to explain the value of basic biomedical research on *Drosophila melanogaster*. Advances in Physiology Education. 35:384–392.

Sasamura, T., Matsuno, K., and Fortini, M. E. (2013) Disruption of *Drosophila melanogaster*. Lipid metabolism genes causes tissue overgrowth associated with altered developmental signaling. PLoS Genet 9(11): e1003917. https://doi.org/10.1371/journal.pgen.1003917.

Spindler S. R. (2005) Rapid and reversible induction of longevity, anticancer and genomic effects of caloric restriction. Mech Ageing Dev. 126(9):960-966. doi: 10.1016/j.mad.2005.03.016. PMID: 15927235.

St. Clair, S. L., Li, H., Ashraf, U., Karty, J. A., and Tennessen, J. M. (2017). Metabolomic analysis reveals that the *Drosophila melanogaster* gene lysine influences diverse aspects of metabolism. Genetics. 207(4): 1255–1261. https://doi.org/10.1534/genetics.117.300201.

Van Timmeren, Steven and Diepenbrock, Lauren and Bertone, Matthew and Burrack, Hannah and Isaacs, Rufus. (2017). A Filter Method for Improved Monitoring of *Drosophila suzukii* (Diptera: *Drosophilidae*) Larvae in Fruit. Journal of Integrated Pest Management. 8. 1-7. 10.1093/jipm/pmx01.

Robertson, J. L., Tsubouchi, A., and Tracey, W. D. (2013). Larval defense against attack from parasitoid wasps requires nociceptive neurons. PloS one, 8(10): e78704. https://doi.org/10.1371/journal.pone.0078704.

Rubin, G. M. and Lewis, E. B. (2000) A brief history of *Drosophila's* contributions to genome research. Science. 287 (5461): 2216-2218.
doi: 10.1126/science.287.5461.2216.

Appendix

Standard Fly Media:

All items can be purchased from Sigma Aldrich and fly media can be purchased ready to use from Archon Scientific. Additional information on where to purchase ingredients and for making your own fly media can also be found at the Indiana University Bloomington *Drosophila* Stock Center website:
https://bdsc.indiana.edu/information/recipes/bloomfood.html

Ingredients:

420 mL water
4.5 gm agar
60 mL of unsulfured molasses
49 gm cornmeal (any kind)
6.5 gm brewer's yeast
145 mL cold water
3.4 mL of propionic acid (acts as a mold inhibitor)

Mix 420 mL of water and agar, bringing the mixture to a boil for about 3–5 minutes. Add unsulfured molasses and heat to boiling again. Mix together cornmeal, brewer's yeast, and cold water in a separate container until all lumps are removed. Add cornmeal-yeast mixture to molasses-agar mixture. Boil mixture for 5 minutes, stirring constantly. Cool mixture to 60°C. Add propionic acid. Pour culture medium 1-inch deep into sterile culture jars with sterile plugs. Add a sprinkle of active baker's yeast (from a saltshaker) to each jar before adding flies.

Additional Diet Options:

To create additional diet options including high fructose, high protein, or high fat diets, soybean extract or coconut oil can be used. These additions can be purchased online or at your local grocery store. These diets can be created using standard media in 5%, 10%, 20%, 40% per weight of diet. Standard fly media food can be used and mixed to these percentages based on wet weight of the readymade food. This is the easiest approach in our experience. It should be noted that to have a higher fat diet than 40% is difficult and can result in suffocation of larvae.

Fly Colony:

Drosophila can be housed in vials partially filled with fly media. The larvae of *D. melanogaster* can be grown in the laboratory within a plastic cage. A plastic beaker and petri dish can be used to create the cage. See: https://www.youtube.com/watch?v=vJ7ZV0hxM5g.

Partially fill Petri dishes with 1% apple juice agar. Take a small amount (~ 1 gram) of the corn meal food and press it against one side of the dish to hold it in place so it will not be dislodged when the Petri dish is turned over when switching out the dishes. A small hole is cut in the bottom of the plastic beaker small enough that a cotton plug can cover the hole. (Note: Use of a soldering iron makes it easy to cut out a hole without the plastic cracking. Don't breathe the fumes from the melting plastic).

A colony can be started by including 50 males and females. Flies can also be ordered through the Indiana University Blooming Stock Center (https://bdsc.indiana.edu) or can be requested from one the author (Robin Cooper). Having adults in this container for 3 days prior to egg pulse will ensure a good amount of freshly collected eggs. Each day for 3 days, preferable first thing in the morning, replace the apple juice agar dish with a fresh dish. On the 3rd day replace the dish for a fresh one and allow the adults to lay eggs for 4 hours. Remove this dish and mark 4-hour egg pulse with time of collection. Repeat again for another 4 hours and label "hours of collection time on the dish and 4- hour egg pulse."

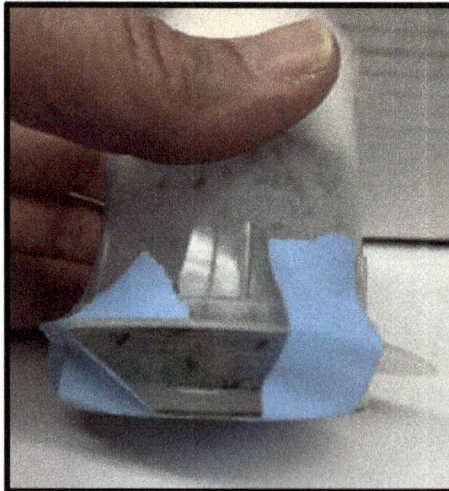

Figure 1.
Example fly colony
with additional plate on bottom.

Use a dish with food to keep the adults alive for the next day so collections can be repeated. The dishes with the eggs cannot have a petri dish lid as the CO_2 will build up and the embryos and larvae will die. We use lids with fine netting glued to the lids and then place the lids on the dishes. The dish can be left for a day like this if one is collecting 1st instars. If one wants to collect 2nd or 3rd instars, then the agar and food will dry out if left without checking for moisture. What works is using a larger petri dish with paper around the smaller dish and a lid placed to the side so as not to trap the CO_2. For additional information on how to knock out flies see:
https://www.youtube.com/watch?v=sGP_5ByY4NM.

After 24 hours at 22-24 degree Celsius the egg dish will start to show the hatching and 1st instars can be collected. These larvae if allowed to continue to develop will yield 2nd instars (day 2), and 3rd instars (day 3). Early 3rd instars will remain in the food and late 3rd instar (day 4) will be wandering and stop eating as they find a place to pupate. Depending on the time to dedicate to this exercise, instars can be removed at the desired stage and placed in the food of choice. Once the colony has started, you can transfer the instars into a long tube containing the experimental food. Continue to repeat this procedure for the different foods to be examined.

Apple juice plates:
Used for colony growth and crawling behaviors

Ingredients:
- 10.1 g Agar (general lab agar)
- 330 mL of water
- 11.1 g of table sugar
- 111 mL of Apple juice (juice, not drink flavor)
- 0.66g *p*-Hydroxybenzoic acid methyl ester, Methyl paraben, NIPAGIN - (SIGMA, catalog # H3647-100G, CAS Number 99-76-3)

Be careful to not breathe this in! Use proper PPE when handling.

Mix agar with water and bring to a boil. Be sure bubbles are occurring to dissolve all the agar. Turn off heat but keep on stirrer. Add sugar and apple juice to mix fully. If one is going to keep the plates for a few weeks add preservative. Pour into the plastic Petrie dish and cover. Then place it in zip lock bag and put in refrigerator.

Microscope:

Watching the larvae crawl and responses to touch, one can use a stereomicroscope (a standard dissecting microscope) with a total of 20x magnification (such as using an eye piece 5x and zoom at 4x). An observer can then record all behavioral responses to touch or number of inch worm movements in a set time period. If one does not have a dissecting scope, one can use simple and inexpensive approaches to magnify the larvae. We made a movie to show various approaches one can use: https://www.youtube.com/watch?v=wTSynwmyHBQ.

Larval Behaviors:

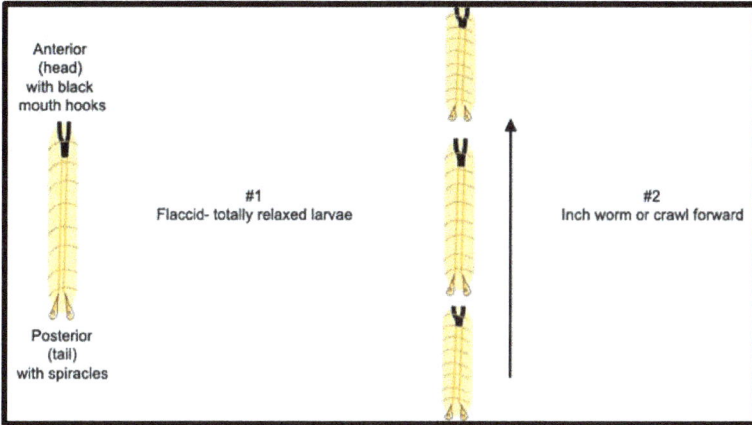

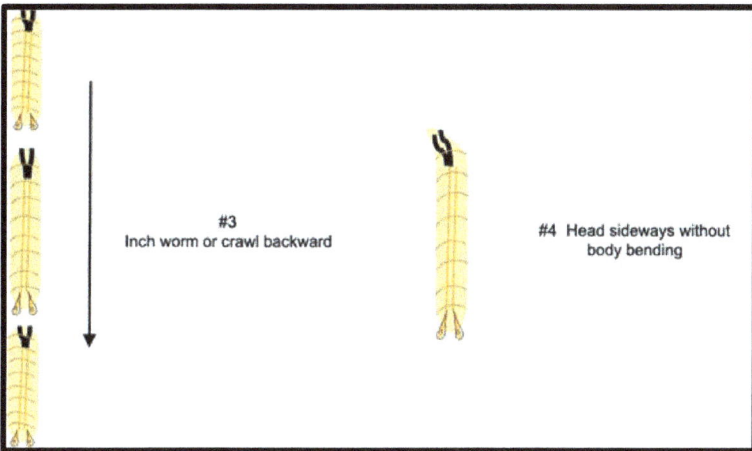

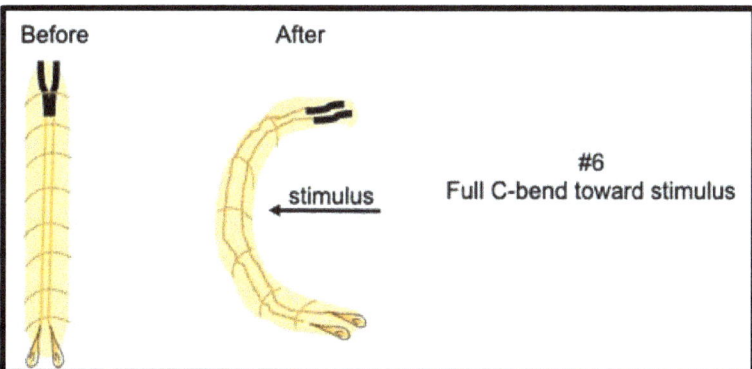

Anterior
(head)
with black
mouth hooks

#1
Flaccid- totally relaxed larvae

#2
Inch worm or crawl forward

Posterior
(tail)
with spiracles

#3
Inch worm or crawl backward

#4 Head sideways without
body bending

Before

After

stimulus

#6
Full C-bend toward stimulus

63

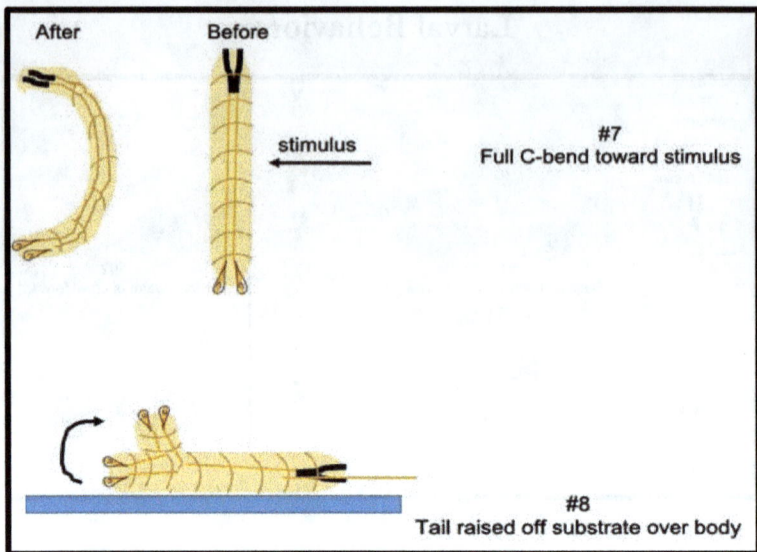

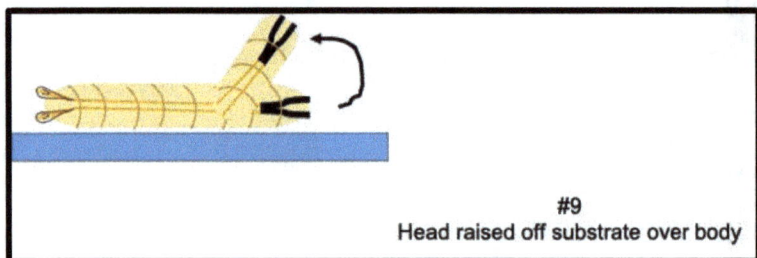

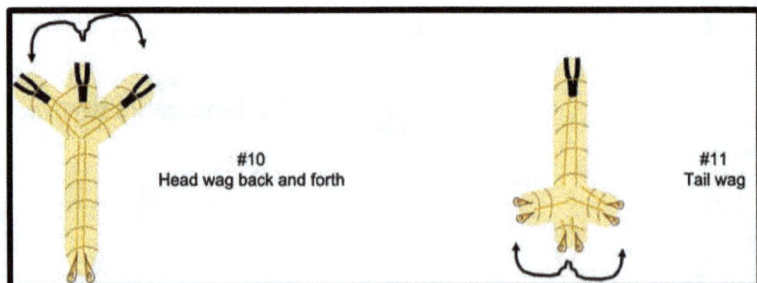

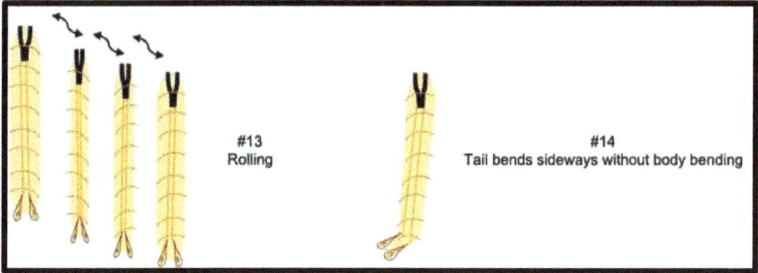

#13
Rolling

#14
Tail bends sideways without body bending

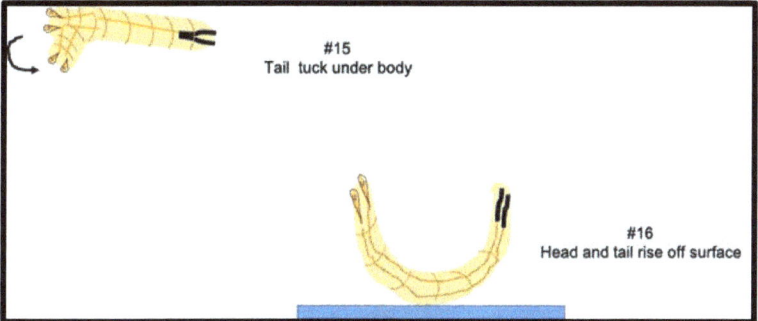

#15
Tail tuck under body

#16
Head and tail rise off surface

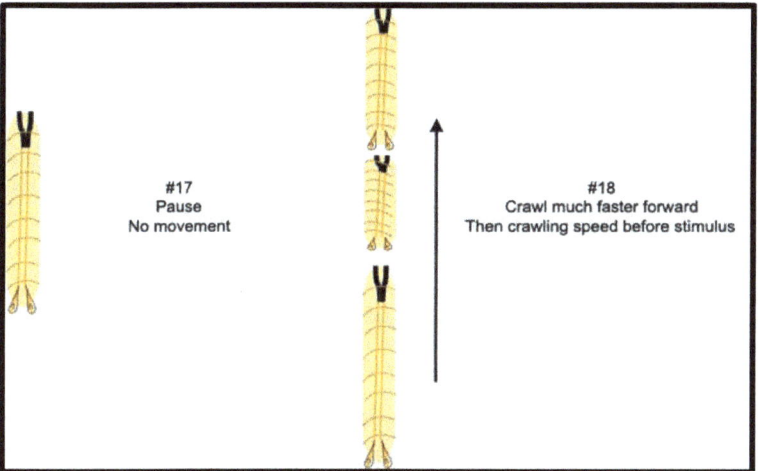

#17
Pause
No movement

#18
Crawl much faster forward
Then crawling speed before stimulus

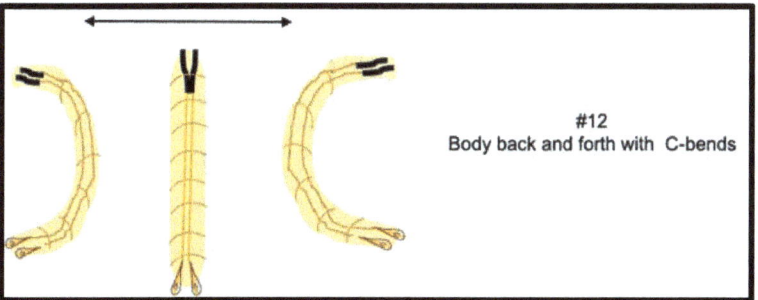

#12
Body back and forth with C-bends

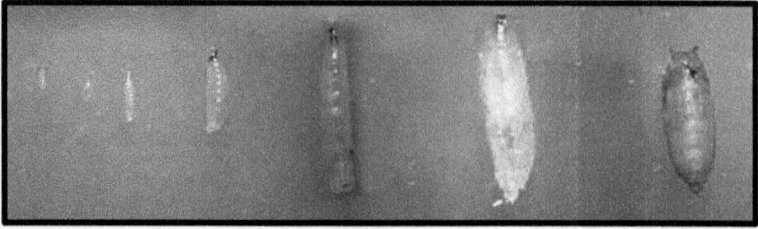

Egg
1st instar,
early 2nd instar, late 2nd instar,
early 3rd instar, late 3rd instar,
pupa

Sample larval behaviors:

1. https://youtu.be/BefxcdAMom8
2. https://youtu.be/mhxnBtDCbRE
3. https://youtu.be/4DFmxsTShgg
4. https://youtu.be/h2CJsHCzdgM
5. No file
6. https://youtu.be/4vwS_J3Locg
7. https://youtu.be/ti3daH6LBEI
8. https://youtu.be/4F69X0JCr5Q
9. https://youtu.be/oQgy6-9WZ_U
10. https://youtu.be/JjiHu7BacXI
11. https://youtu.be/PUEshv0Aw7E
12. https://youtu.be/OUZm7zdyKzU
13. https://youtu.be/dHIbafY0b34
14. https://youtu.be/SgPX3R0n1yA
15. https://youtu.be/YrCMLq3gfFc
16. https://youtu.be/8CzL2SjmWNI
17. https://youtu.be/QRQoNlSn7Xw

Acknowledgments

Students over the years helped develop these aspects of this module in various ways:

LaShay Byrd, Jenni Ho, Ruth Sifuma, Brecken Overly, Hunter Maxwell and Crysta Meekins, Brittany Slabach, Alex Stanback, Maddie Stanback, Emma Rotkis, Clare Cole, Kay Johnson, Samantha Danyi, Sushovan Dixit, Madan Subheeswar, Gaayathri Veeraragavan, Suraj Rama, Angel Ho, and Samuel Potter.

Teachers who have helped develop the content are Kim Zeidler-Watters and Diane Johnson.

Chapter 8:

Forensics for the body farm: Preferences for the medicinal blow fly (*Phaenicia sericata*) and fruit fly (*Drosophila melanogaster*).

Level:
High school and college

Aspects of this protocol were previously presented as a poster. A more detailed module is presented below.

Oscar Istas, Abigail Greenhalgh, Erin E. Richard, Jate Bernard, Rebecca Krall, Tawny Aguayo, and Robin L. Cooper. 2022. Volume 42–Istas supplement. In: McMahon K., editor. Publication of the Association for Biology Laboratory Education. Volume 42, Article 61 2022. http://www.ableweb.org/volumes/vol-41/?art=#.

https://www.ableweb.org/biologylabs/wp-content/uploads/volumes/vol-42/61_Istas.pdf

Overview

Learning about bacteria, fungi, or the developmental stages of insects does not always have the "wow" factor for many college students. If you add a dead body to the mix, it's amazing how their interest is piqued.

An interactive forensic science module was developed with this storyline to provide an authentic forensic investigation of a dead animal. Forensic science considers many variables when documenting and determining conditions to identify the potential time frame of death. In particular, the presence, amount, and developmental stages of bacteria, fungi and insects are commonly used to aid investigations. The developmental stages and preference in behaviors of insects, and the decay of associated plant matter has proven to be particularly beneficial in determining the potential time frame of death of an animal.

The case in this module presents a human that has died after taking a bite of fruit. Fruit flies and blow flies are found at the scene. Through experimentation, data gathering, and analysis of the life cycles and behavior of two animal models (fruit fly and blow fly), interpretations of the location of the insects and developmental stages of larvae and pupa lead to a logical assessment of the time of death.

We simulated the experimental design of data collection at the scene using laminated copies of fruit fly and blow fly larvae in different developmental stages. The data set and specific details surrounding data collection are provided to support participants in determining the time frame of the dead animal using an experimental protocol. Protocols also are provided to guide the re-creation of the scene using physical models. The lab has been designed to be conducted in the laboratory or remotely using downloadable materials. The laboratory can also be adapted as a CURE class project.

Introduction:

The ability to solve problems given various content is rewarding for students (National Research Council, 2012). Curiosity can be piqued by the successful applications of data collection and analysis to an engaging topic, especially if multiple correct conclusions can be reached. In considering the strong interest in crime science investigations (CSI) currently in society, having an educational model to feed this interest could engage students familiar with the general topic. Bringing in forensic science is a means to introduce scientific investigation and complexity of the field. The diverse array of factors which contribute to forensic investigation allow the development of thinking on the implications of the environment in solving a problem. This format allows instructors to diversify instruction in a classroom each time the content is taught.

In this module, we set up a template for CSI by introducing elements of entomology, microbiology, and environmental science in order to work through a relatively superficial problem. As one dives deeper into the content, the more complex the investigations can become. This module is structured in a way to encourage further investigation in different avenues depending on the goals of the participants and instructors.

The scenario is that a human body (which could be substituted with a different mammal: a fruit eating bat or primate) was found in an open field with a bite taken out of an apple next to the body. The environment is set as a mild weather day (dry, 21°C), but can easily be varied. In this exercise we focus on two insect species. The first is a common fruit fly (*Drosophila melanogaster*), which is found in the apple (which has a broken peel). This is a factor to be studied as *Drosophila melanogaster* will not lay eggs on an intact apple. The other insect is a blow fly, commonly known as the green fly, which consumes and lays eggs on decaying flesh from dead animals such as roadkill (i.e., deer, squirrels). The eggs and pupa of these two types of flies are unique in morphology and develop at different rates depending on the temperature. The apple and body can be identified with or without bacteria and fungi present to add a variable to examine the impact on the development and survival of the insects. The main focus is determining the time it would take to see empty pupal cases of the two fly species to estimate the time of death of the subject.

The investigation can be simulated with kits containing pictures of the flesh and apple with different developmental stages of the flies as well as with or without bacteria or fungi. The developmental stages of the insects are illustrated with a diagram of the stages and the time required to reach each of the stages. An instructor might put more empty pupa cases of one insect species as compared to the other to stimulate discussion about the behavior and egg-laying preference of the insects. Setting the temperature to a standard temperature of 21°C in a dry environment would provide a quick determination of a window of time that the apple was bitten and dropped, considering *Drosophila* found the apple quickly and laid eggs. If an empty pupa case is found, then back estimation is about 8 to 9 days. However, the development is temperature dependent. This will allow participants to investigate how to obtain environment conditions from web-based sources of their own local environment or set dates and location an instructor chooses. The presence and types of bacteria and fungus can kill the embryos in the eggs as well as larva and pupa, which can alter the estimation in the time of occurrence of death.

In recreating the scene, laboratory investigations can be implemented with the insects on fruit and tissue such as beef liver or uncooked pieces of meat in either an indoor lab or an outdoor field setting. Indoors offers controlled conditions such as temperature, lighting, and humidity, whereas outdoors can be more variable and would require more investigation of environmental conditions during the time frame. To refine indoor investigations, plastic plates with tainted items (apple, meat) and inoculated bacteria or fungi can be implemented. This could be even more refined to include variation in toxins released by a given bacterial species such as lipopolysaccharides (LPS) by gram negative bacteria. Similarly, providing adult insects a choice in feeding and egg-laying material can encourage investigation into behavioral choices and impact with bacterial toxins on food.

We provide information including the development stages of both insect species which can be accessed by students online or cut out, placed in plastic bags, and mailed to students for hands-on remote learning. This content could also be emailed to students and encourage them to set up their own conditions for the investigation.

Student Outline

Objectives:
Students will be able to:

1. Integrate observations to make predictions based on evidence.
2. Utilize literature research to help make a prediction.
3. Identify developmental stages and environmental impacts for the insects used in the study.
4. Create and design a model to support the evidence
5. Describe how forensic scientists use evidence and inference to solve a problem.
6. Discuss diet choices and impact of variables such as the presence of bacteria and their toxins for insect larvae.

Introduction:
The premise of this module is a hypothetical crime scene where someone has died in a rural farm field. When the forensic team arrived, they noticed the man appeared to have taken a bite out of an apple sometime before his death, which was next to the body. They also noticed adult flies of the species *Drosophila melanogaster* and blow flies

flying around the body and the apple. Photos were taken of the apple and the person where the flies were aggregating.

The instructor of the module may provide varied data and details of the environment to present different scenarios to groups within a class or for different years in teaching this module. Below is the data we provide as a template for potential factors which could be relevant to determining the time of death of the body. Further investigation into the location and developmental stages of the insects will shed light on the matter.

The goal is for you is to estimate the time frame in which the person died with or without additional experimentation. Protocols are provided to recreate the scene with physical modeling. Variations in the experimentation are detailed with agar plates and food for insect developmental studies. There are various tasks that one might be assigned to work through.

Conditions:

For the last 2 weeks, the weather has consisted of mild weather 70°F (21.1°C) during the day and at nighttime down to 55°F (12.8°C). The location is a horse farm in central Kentucky, USA in an open grassy field. No rain for the last two weeks was reported. Some gram-negative bacteria (*Serratia marcescens*) was present on the apple, along with a little bit of fungus.

The photo of the area is provided.

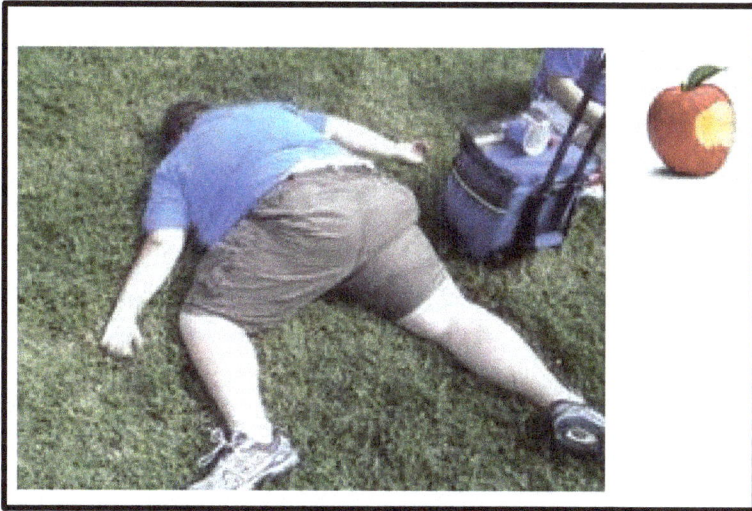

Figure 1:
The body and apple in the location found.

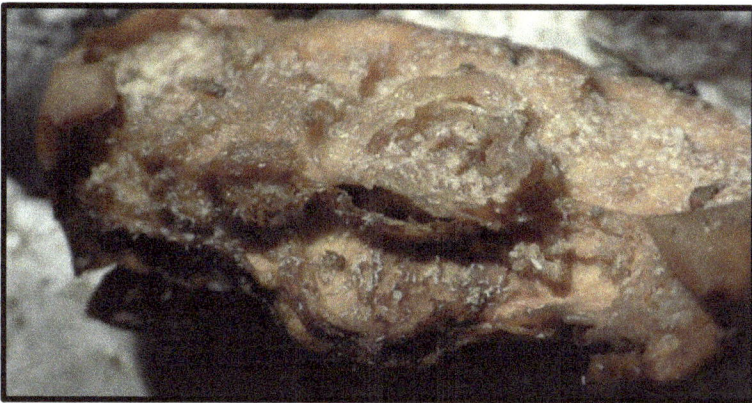

Figure 2:
A close up photo of the decaying apple
with egg cases and larvae of all stages and some pupa.
Mostly *Drosophila* stages are present
but with a few of the blow fly.

Figure 3:
Several empty pupa cases of *Drosophila*
were found on the ground rather than on the apple.

Figure 4:
Simulated photo of the tongue
of body (in practice this is of beef liver).
Dead blow flies and empty pupa cases
of blow flies are noted.

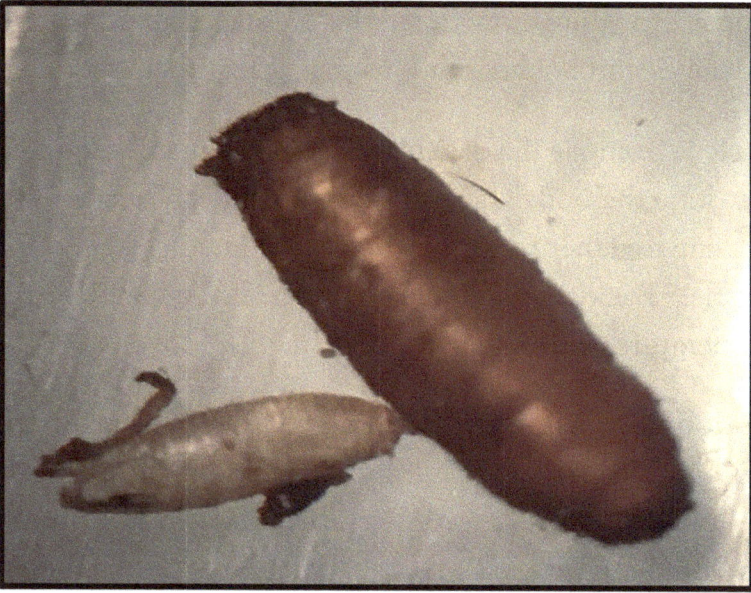

Figure 5:
**These pupae are found in the grass
between the apple and the subject's face.
The larger one is of the blow fly.**

The activity is divided into flexible tasks for teachers to adjust as needed but to provide a logical framework for stepwise learning.

Task 1: Look over the data provided and come up with a list of information about the insects that could help to estimate the time in which the animal/person might have died. Students make lists and then write them out on a board and see how many of the same items were chosen.

Task 2: Examine the literature and web sources to find the life cycle of fruit flies and blow flies and how this relates to forensics. Topics to pay attention to: Temperature, food source, crowding, how to tell the life stages apart, length of life stages and how conditions may affect them.

Task 3: Compile the data provided to estimate at minimum how long the body and fruit must have been present. Put a timeline together based on the developmental stages of the insects. Back date down to egg laying and list air temperatures with the dates (day/night).

Optional—Task 4: Set up a simulation with a cut apple and beef liver. Add fruit flies and blow flies. Conduct experiments at room temperature and monitor developmental stages.

Optional—Task 5: Set up isolated fruit flies and blow files in separate containers with food. Use incubators to simulate temperature changes and monitor developmental stages.

Details for each task:

Task 1: Potential variables students may consider: Body temperature, condition of corpse (skin broken or intact), insect larvae inside body or only around mouth and eyes, leaking body fluids, dehydration, hair falling out, grass/plants underneath dead or look fresh and green, insects under body, wild animal bites from dogs or other large animals, insects associated with body. If insects present, what stages?

Apple: dried out or moist, bacteria, fungi, insects present. If insects are present, what stages?
Environment: Temperature of the last few days, precipitation, wind.

Task 2: Google searches on:
- Life cycle *Drosophila*
- Life cycle blow flies (*Phaenicia sericata*)
- How to stage larvae, temperature effects on insect development
- How to determine how long an animal is dead, forensics dead animal, forensics insects.

Task 3: List out the stages of the two different types of larvae, eggs, pupa, and if pupa cases are enclosed. Try to make a developmental curve based on temperatures. Back calculate the potential dates that the person and apple were exposed to the open environment. Use Netlogo simulation to examine how fast a population can grow depending on number of adults and sex of adults.

See the information on these hot links.
One needs to download the free netlogo software for these modules to function.

https://ccl.northwestern.edu/netlogo/download.shtml
Module 1 . . . Module 2 . . . Module 3

Task 4: Go over how to simulate the scene with mixed fly species and how to monitor the food sources and insect development. For an outside simulation, one needs a cage to keep other animals from running off with content. If the cage is small enough then it can be brought indoors. If one wanted to simulate outdoor conditions, one could use a semi closed or could use chicken wire to make it open for flies but to keep large animals away. This would be a more natural condition.

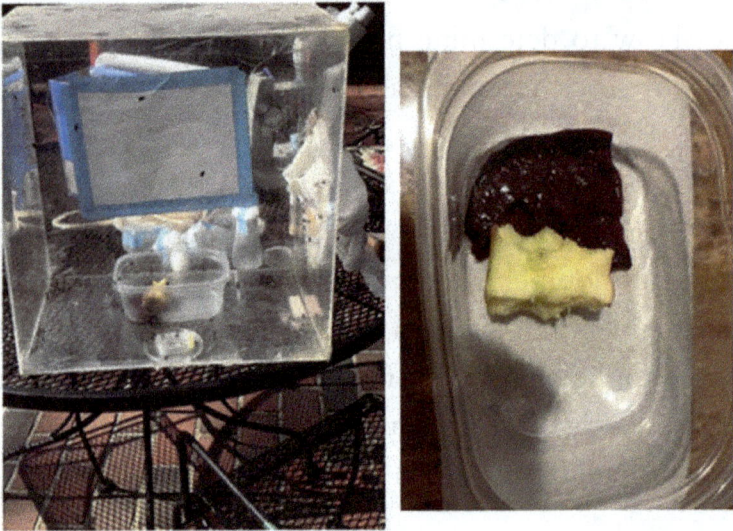

Figure 6:
**A plastic cage with screening to allow air in and out.
Apple and beef liver placed inside
along with adult *Drosophila* and blow flies**

Task 5: Go over how to investigate individual fly species and effects of the environment on the developmental cycle. One can make agar plates with LPS within the agar or food and at different concentrations. *D. melanogaster* avoid eating food containing bacterial LPS. This gustatory avoidance was shown to be mediated through a TRPA1 receptor (Soldano, et. al., 2016).

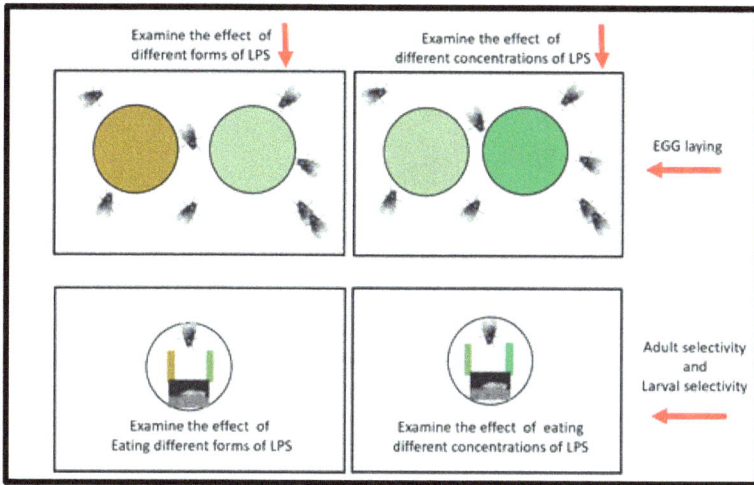

Figure 7:
A potential experimental design to investigate the choices of blow flies and *Drosophila* for egg laying depending on the presence and concentration of LPS in the agar (top panel).

The choice of eating food not tainted or tainted with LPS and variation in concentration of LPS. Here food is placed close together in strips so larvae could choose relatively easily between the food groups after emerging from the egg cases.

Discussion:

Now that one has a snapshot of the conditions in which the body and apple were found, one should be able to determine how long (approximate days or hours) the body and the apple have been in the environment.

Depending on the tasks assigned to you, write out a description of your notes from each task as if presenting a report to a group of fellow forensic scientists. Provide details on how the outcome of how each task was managed. In explaining your results, explain the steps that led you to this outcome, as well as any potential confounding factors.

Cited References:

Soldano, A., Alpizar, Y. A., Boonen, B., Franco, L., López-Requena, A., Liu, G., Mora, N., Yaksi, E., Voets, T., Vennekens, R., Hassan, B. A., Talavera K. 2016. Gustatory-mediated avoidance of bacterial lipopolysaccharides via TRPA1 activation in *Drosophila*. Elife. 5:e13133.

Materials:

One bottle of larvae that has been stored at room temperature.

Apple and beef liver if recreating the module.

Wire cage for outdoor/indoor recreation of conditions (use nylon stocking for netting screen or chicken wire)

2 medium-size Petri dishes (for feeding experiments)

Small paint brush

Cornmeal food for mixing with LPS

LPS from Sigma-Aldrich chemical Co.

Dissecting microscope or phone camera

For remote learning (kit or download files)

Notes for the Instructor:

The challenging aspect of conducting a live animal project is being able to monitor the setup if placed outside, due to interference from wild animals. If held indoors, the potential for flies to escape the plastic chamber is a possibility. This project is best conducted in the season where blow flies are easily recruited with bait left outside. In addition, fruit flies are readily be obtained on warm days in many locations; however, they can also be obtained by contacting most university researchers who work with fruit flies, or by ordering them from a stock center. So if you want outsides flies to go to the food sources then use chicken wire around a food items to keep raccoons, squirrels and birds out. If one wants to maintain flies one puts in the cage but need air to flow, then use material from nylon stockings to cover the holes.

It may be advantageous to have the students write a report of their findings and how their conclusions were reached.

Students should find that the body and fruit have been present at the scene of death for [x] days, based on the temperature, weather, and life stages of the fruit flies and blow flies. Students should explain the steps that led them to this outcome, as well as any potential confounding factors. Their task is not to draw conclusions about the cause of death, only to use various sources of information to deduce a reasonable time frame. Encourage discussion about different factors that may cause the body and fruit to decompose and promote insect development at different rates.

If an instructor wanted to use their local environments or place a date and location for the scene, then students could look up weather information online or examine ways to best estimate the weather conditions during the time frame.

Examining how flies choose between food that is tainted with LPS can stimulate discussion about what types of bacteria are in the environment and how these species differ. Different bacterial strains of LPS can be obtained from Sigma-Aldrich chemical company. There are likely some forms of LPS which will generate unique results and could be used for reporting novel findings as there is not an exhaustive amount of published research on this topic. Publications in undergraduate or primary research journals are a possibility.

The use of LPS instead of bacteria offers an easier environment to control than having to grow bacteria and worry about identification of species. It would be best if the instructor of the course was the one to mix the food and LPS as well as make agar dishes with LPS as the LPS needs to be weighed out in a hood with proper protective gear so as not to ingest or inhale the powder.

There are many variations to this module which can be altered each time it is taught. The environmental conditions can be changed as well as the type of insects used. Pill bugs and other common insects can be used to examine food preference and effects of LPS forms.

For remote learning with participants please download a kit for sending to students after adjusting the content as needed.

http://web.as.uky.edu/Biology/faculty/cooper/ABLE-2021/ABLE-2021-Body%20farm/Home-Forensics%20for%20the%20body%20farm-ABLE%202021.htm

Cited Reference:

National Research Council. 2012. *A Framework for K-12 Science Education: Practices, Crosscutting Concepts, and Core Ideas.* (2012). Committee on a Conceptual Framework for New K-12 Science Education Standards. Board on Science Education, Division of Behavioral and Social Sciences and Education. Washington, DC: The National Academies Press.

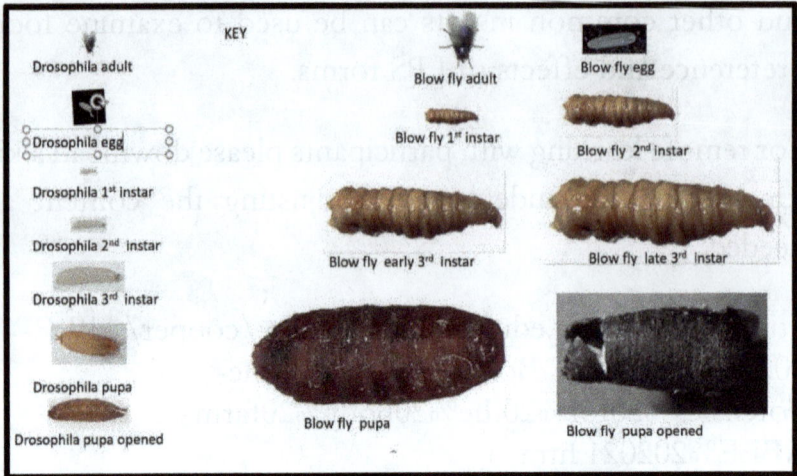

KEY

Drosophila adult

Drosophila egg

Drosophila 1st instar

Drosophila 2nd instar

Drosophila 3rd instar

Drosophila pupa

Drosophila pupa opened

Blow fly adult

Blow fly 1st instar

Blow fly early 3rd instar

Blow fly pupa

Blow fly egg

Blow fly 2nd instar

Blow fly late 3rd instar

Blow fly pupa opened

**Figures that can be copied
and put into kits for distant learning
or individual learning.**

Chapter 9:

Developing CURE/ACURE projects
for publishing as a research article

A number of class projects for freshman and senior college classes turned out to be fruitful enough to continue the projects as research projects to see through to publication. The four themes that turned out well are presented below as examples of class projects that we have used. This is why we define ACURE as Authentic Curriculum Undergraduate Research Experience. They were as follows:

Example 1:
Examining the effects of exposure to excess essential metals and pharmacological agents on various tissues in invertebrate models.

Example 2:
Actions of doxapram used in a class module led to primary research projects.

Example 3:
Effects of Ca^{2+} on sensory neurons.

Example 4:
From the fly body farm teaching module to research focus on effects of bacterial lipopolysaccharides.

Each of these examples is presented as the tangibles that were produced from the projects.

Example 1:
Examining the effects of exposure to excess essential metals and pharmacological agents on various tissues in invertebrate models.

One of the themes on the effects of excess essential metals and supplements from health food suppliers. The class projects were to determine what excessive amounts of the essential metals may have on physiological function and overall survival using invertebrate models.

Below are listed the publications we developed with this focus:

Dayaram, V., Malloy, C., Martha, S., Alvarez, B., Chukwudolue, I., Dabbain, N., D.mahmood, D., Goleva, S., Hickey, T., Ho, A., King, M., Kington, P., Mattingly, M., Potter, S., Simpson, L., Spence, A., Uradu, H., Van Doorn, J. L., and Cooper, R. L. (2017). Stretch activated channels in proprioceptive chordotonal organs of crab and crayfish are sensitive to Gd3+ but not amiloride, ruthenium red or low pH. IMPLUSE The Premier Undergraduate Neuroscience Journal. https://impulse.appstate.edu/issues/2017.

Malloy, C., Dayaram, V., Martha, S., Alvarez, B., Chukwudolue, I., Dabbain, N., D.mahmood, D., Goleva, S., Hickey, T., Ho, A., King, M., Kington, P., Mattingly,

M., Potter, S., Simpson, L., Spence, A., Uradu, H., Van Doorn, J. L., Weineck, K. and Cooper, R. L. (2017). The effects of neighboring muscle injury on proprioception responses in crayfish and crab. J. of Exp. Zoology. 327(6):366–379.

Dayaram, V., Malloy, C., Martha, S., Alvarez, B., Chukwudolue, I., Dabbain, N., D.mahmood, D., Goleva, S., Hickey, T., Ho, A., King, M., Kington, P., Mattingly, M., Potter, S., Simpson, L., Spence, A., Uradu, H., Van Doorn, J. L., and Cooper, R. L. (2017). The effect of CO_2, intracellular pH and extracellular pH on mechanosensory proprioceptor responses in crayfish and crab. American Journal of Undergraduate Research 14(3):85-99.

Atkins, D. E., Bosh, K. L., Breakfield, G. W., Daniels, S. E, Devore, M. J., Fite, H. E., Guo, L., Henry, D., Kaffenberger, A., Manning, K. S., Mowery, T., Pankau, C. L., Parker, N., Serrano, M. E., Shakhashiro, Y., Tanner, H., Ward R. A., Wehry, A. H., and Cooper, R. L. (2021) The effect of calcium ions on mechanosensation and neuronal activity in proprioceptive neurons. NeuroSci 2021, *2*: 353-371. https://doi.org/10.3390/neurosci2040026 or https://www.mdpi.com/2673-4087/2/4/26/htm.

Pankau, C., Nadolski, J., Tanner, H., Cryer, C., Di Girolamo, J., Haddad, C., Lanning, M., Miller, M., Neely, D., Wilson, R. Whittinghill, B. and Cooper, R. L. (2022) Effects of manganese on physiological processes in *Drosophila*, crab and crayfish: Cardiac, neural and behavioral assays. Comparative Biochemistry and

Physiology Part C. vol. 251, 2022, 109209. ISSN 1532-0456, https://doi.org/10.1016/j.cbpc.2021.109209.

Pankau, C. and Cooper, R. L. (2022) Molecular physiology of manganese in insects. Current Opinion in Insect Science. 2022, 51:100886.

Wagers, M., Starks, A., Abul-Khoudoud, M. O., Ahmed, S. M., Alhamdani, A. W., Ashley, C., Bidros, P. C., Bledsoe, C. O., Bolton, K. E., Capili, J. G., Henning, J. N., Ison, B. J., Moon, M., Phe, P., Stonecipher, S. B., Tanner, H. N., Taylor, I. N., Turner, L. T., West, A. K. and Cooper, R. L. (2023). An invertebrate model to examine the effect of acute ferric iron exposure on proprioceptive neurons. Comparative Biochemistry and Physiology Part C. 266: 109558, https://doi.org/10.1016/j.cbpc.2023.109558 (Bio 446/650 class project).

Wagers, M., Starks, A., Nadolski, J., Bierbower, S. M., Altenburg, S., Schryer, B. and Cooper, R. L. (2024) Examining the effect of iron (ferric) on physiological processes: Invertebrate models. Comparative Biochemistry and Physiology- Part C. 278 (2024) 109856. https://doi.org/10.1016/j.cbpc.2024.109856 .

Elliott, E. R., Nadolski, J., McIntosh, R. D., Datta, M.S., Crawford, D. M., Hirtle, J. T., Leach, A. B., Roemer, K.A., Sotingeanu, L. C., Taul, A. C., Vessels, B. D., Speed, S. L., Bradley, A. L., Farmer, C. G., Altenburg, S., Bierbower, S. M. and Cooper, R. L. (2023) Investigation regarding the physiological effects of zinc on *Drosophila* and crawfish cardiac, neural, synaptic, and behavioral processes. Journal of Pharmacology and Toxicology. 18: 166-189.

Brock, K. E., Elliott, E. R., Taul, A.C., Asadipooya, A., Bocook, D., Burnette, T., Chauhan, I.V., Chhadh, B., Crane, R., Glover, A., Griffith, J., Hudson, J. A., Kashif, H., Nwadialo, S. O., Neely, D. M., Nukic, A., Patel, D., Ruschman, G. L., Sales, J. C., Yarbrough, T., and Cooper, R. L. (2023). The effects of lithium on proprioceptive sensory function and nerve conduction. NeuroSci.4: 280-295 (Bio446/650 class project). https://www.mdpi.com/2673-4087/4/4/23.

Elliott, E. R., Brock, K. E., Taul, A. C., Asadipooya, A., Bocook, D., Burnette, T., Chauhan, I. V., Chhadh, B., Crane, R., Glover, A., Griffith, J., Hudson, J. A., Kashif, H., Nwadialo, S.O., Neely, D. M., Nukic, A., Patel, D., Ruschman, G. L., Sales, J.C., Yarbrough, T., and Cooper, R. L. (2023) The effects of zinc on proprioceptive sensory function and nerve conduction. NeuroSci 4(4), 305-318; (Bio446/650 class project). https://www.mdpi.com/2673-4087/4/4/25.

Brock, K. E., Gard, J. E., Elliott, E. R., Taul, A. C., Nadolski, J., Kim, J., McCubbin, S., Thuy, A., Ronen, R., Bierbower, S .M., Alcorn, J. P., Dharanipragada, N., Hall, T. F., Hamlet, A. B., Iqbal, Z., Johnson, S. R., Joshi, J. K., McComis, S. J., Neeley, R. E., Racheneur, A. W., Satish, D., Simpson, T. R., Walp, J. L., Murray C., Wright, J. and Cooper, R. L. (2024) Investigation regarding the physiological effects of sodium selenite on physiological functions in *Drosophila*, crayfish, and crab: behavior, cardiac, neural, and synaptic properties. In press, Journal of Pharmacology and Toxicology.

Gard, J. E., Brock, K. E., Elliott, E. R., Taul, A. C., Nadolski, J., Kim, J., McCubbin, S., Thuy, A., Ronen, R., Bierbower, S. M., Alcorn, J. P., Dharanipragada, N., Hall, T. F., Hamlet, A. B., Iqbal, Z., Johnson, S. R., Joshi, J. K., McComis, S. J., Neeley, R. E., Racheneur, A. W., Satish, D., Simpson, T. R., Walp, J. L., Murray C., Wright, J. and Cooper, R. L. (2024) Investigation regarding the physiological effects of cobalt on physiological functions in *Drosophila*, crayfish and crab: behavior, cardiac, neural, and synaptic properties. Comparative Biochemistry and Physiology Part C: Toxicology and Pharmacology, 2025, 110165. https://doi.org/10.1016/j.cbpc.2025.110165.

Some of these past studies led to the students also wanting to address various pharmacological agents on neurons and physiological aspects on these invertebrate models.

During this next project a clinical neurologist zoomed in to our class to explain why Riluzole is used for people with Amyotrophic Lateral Sclerosis (ALS).

Nethery, B., Abou El-Ezz, M., Brown, C., Calderaro, T., Evans, C., Grant, T., Hazelett, R., High, C., Buendia Castillo, D., Ilagan, T., Klier, J., Marguerite, N., Marino, F., McCubbin, S., Meredith, N., Naidugari, P., Russell, W., Sommers, N., and Cooper, R. L. (2021) The effects of Riluzole on sensory and motor nerve function. IMPULSE. This was a Bio446 class project for Fall 2020. https://impulse.appstate.edu/articles/2021/effects-riluzole-sensory-and-motor-nerve-function.

Brock, K. E., Gard, J. E., Elliott, E. R., Taul, A. C., Nadolski, J., Kim, J., McCubbin, S., Thuy, A., Ronen, R., Bierbower, S. M., Alcorn, J. P., Dharanipragada, N., Hall, T. F., Hamlet, A. B., Iqbal, Z., Johnson, S. R., Joshi, J. K., McComis, S. J., Neeley, R.E., Racheneur, A. W., Satish, D., Simpson, T. R., Walp, J. L., Murray C., Wright, J. and Cooper, R. L. (2025) Investigation regarding the physiological effects of sodium selenite on physiological functions in *Drosophila*, crayfish, and crab: behavior, cardiac, neural, and synaptic properties. (Bio446 class project). Journal of Pharmacology and Toxicology 20:1-20 [PDF]. https://scialert.net/fulltext/fulltextpdf.php?pdf=academicjournals/jpt/2025/1-20.pdf.

Gard, J. E., Brock, K. E., Elliott, E. R., Taul, A. C., Nadolski, J., Kim, J., McCubbin, S., Thuy, A., Ronen, R., Bierbower, S. M., Alcorn, J. P., Dharanipragada, N., Hall, T. F., Hamlet, A. B., Iqbal, Z., Johnson, S. R., Joshi, J. K., McComis, S. J., Neeley, R. E., Racheneur, A. W., Satish, D., Simpson, T. R., Walp, J. L., Murray C., Wright, J. and Cooper, R. L. (2025) Investigation regarding the physiological effects of cobalt on physiological functions in *Drosophila*, crayfish and crab: behavior, cardiac, neural, and synaptic properties. (Bio446 class project). Investigation regarding the physiological effects of cobalt on physiological functions in *Drosophila*, crayfish and crab: behavioral, cardiac, neural, and synaptic properties. Comparative Biochemistry and Physiology Part C: Toxicology and Pharmacology, 2025, 110165, https://doi.org/10.1016/j.cbpc.2025.110165. (Impact factor: 3.9).

This next class project focused on why people used 4-AP to treat or mediate the effects of MS which led to this manuscript and then an educational module to follow for high school or college level classes.

Tanner, H., Atkins, D. E., Bosh, K. L., Breakfield, G. W., Daniels, S. E, Devore, M. J., Fite, H. E., Guo, L., Henry, D., Kaffenberger, A., Manning, K.S., Mowery, T., Pankau, C. L., Serrano, M. E., Shakhashiro, Y., Ward R. A., Wehry, A. H., and Cooper, R. L. (2022) The effect of TEA and 4-AP on primary sensory neurons in a crustacean model. Journal of Pharmacology and Toxicology. 17: 14-27.

Exploring mechanisms in a medical treatment for a disease: A teaching/learning module.

Published—so as an example for a college level course with access to electrophysiological instrumentation.

O'Neil, A. S., Krall, R. M., Vascassenno, R., Cooper, R. L. 2023. Exploring mechanisms in a medical treatment for a disease: A teaching/learning module. Article 35 In: Boone E. and Thuecks S., eds. Advances in biology laboratory education. Volume 43. Publication of the 43rd Conference of the Association for Biology Laboratory Education (ABLE).
https://doi.org/10.37590/able.v43.art35.

https://web.as.uky.edu/Biology/faculty/cooper/labW WW-PDFs/ABLE%202023%20paper.pdf.

Example 2:
Actions of doxapram used in a class module led to primary research projects. In this project case a Clinical PharmD and a anesthesiology remotely talked to the class, via Zoom, to explain what Doxapram is for used clinically.

Ison, B. J., Abul-Khoudoud, M. O., Ahmed, S. M., Alhamdani, A. W., Ashley, C., Bidros, P. C., Bledsoe, C. O., Bolton, K. E., Capili, J. G., Henning, J. N., Moon, M., Phe, P., Stonecipher, S. B., Tanner, H. N., Taylor, I. N., Turner, L. T., Wagers, M., West, A. K. and Cooper, R. L. (2022). The effect of Doxapram, a K2p channel blocker, on proprioceptive neurons: Invertebrate model. NeuroSci 2022, 3, 566-588. https://doi.org/10.3390/neurosci3040041.

This past paper led to these next couple of papers:

Vacassenno, R. M., Haddad, C. N. and Cooper, R. L. (2023) The effects on resting membrane potential and synaptic transmission by Doxapram (blocker of K2p channels) at the *Drosophila* neuromuscular junction. Comparative Biochemistry and Physiology Part C 263 (2023) 109497. https://www.sciencedirect.com/science/article/abs/pii/S1532045622002320#ab0015.

Vacassenno, R. M., Haddad, C. N. and Cooper, R. L. (2023) Bacterial lipopolysaccharide hyperpolarizes the membrane potential and is antagonized by the K2p channel blocker doxapram. Comparative Biochemistry and Physiology Part C. 266: 2023, 109571, https://doi.org/10.1016/j.cbpc.2023.109571.

Elliott, E. R., Brock, K. E., Vacassenno, R. M., Harrison, D.A., Cooper, R. L. (2024) The effects of doxapram and its potential interactions with K2P channels in experimental model preparations. Journal of Comparative Physiology A, 2024. https://doi.org/10.1007/s00359-024-01705-6.

Example 3:

Effects of Ca2+ on sensory neurons. This class project led to the primary research focus on how altering extracellular Ca2+ concentrations altered nerve excitability.

Atkins, D. E., Bosh, K. L., Breakfield, G. W., Daniels, S. E, Devore, M. J., Fite, H. E., Guo, L., Henry, D., Kaffenberger, A., Manning, K. S., Mowery, T., Pankau, C. L., Parker, N., Serrano, M. E., Shakhashiro, Y., Tanner, H., Ward R. A., Wehry, A. H., and Cooper, R. L. (2021) The effect of calcium ions on mechanosensation and neuronal activity in proprioceptive neurons. NeuroSci 2021, 2: 353-371.
https://doi.org/10.3390/neurosci2040026
or https://www.mdpi.com/2673-4087/2/4/26/htm.

This first study led to this next primary paper:

Elliott, E. R. and Cooper, R. L. (2024). The effect of calcium ions on resting membrane potential. BIOLOGY. 2024, 13, 750.
https://doi.org/10.3390/biology13090750

Example 4:

From the fly body farm teaching module to research focus on effects of bacterial lipopolysaccharides.

As mentioned above, one theme on why different flies (larvae and adults) can live and survive in food laced with bacteria started with a simple class project which in time turned into a major research focus over the next few years.

The module on body farm for high school/college (Chapter 8) led to a primary paper and many aspects to follow up on. This is a good example of playing around and then following up on observations and turning the CURE project into ACURE projects.

Stanback, M., Stanback, A. E., Akhtar, S., Basham, R., Chithrala, B., Collis, B., Heberle, B. A., Higgins, E., Lane, A., Marella, S., Ponder, M., Raichur, P., Silverstein, A., Stanley, C., Vela, K. and Cooper, R. L. (2019). The effect of lipopolysaccharides on primary sensory neurons in crustacean models. IMPLUSE.
https://impulse.appstate.edu/articles/2019/effect-lipopolysaccharides-primary-sensory-neurons-crustacean-models.

This led to two more educational projects and a review of how to conduct these semester long ACURE projects:

Bernard, J., Danley, M., Krall, R., Sharp, K., Cooper, R. L. (2022). Authentic curriculum undergraduate research experimentation to learn about the effects of septicemia on cardiac function: frog and larval *Drosophila* models.

Article 53. In: Advances in Biology Laboratory Education. Volume 42. Proceedings of the 41st Conference of the Association for Biology Laboratory Education (ABLE).

Sharp, K. A., Krall, R. M., Cooper, R. L., Danley, M., Barnard, J. (2022). What do animal physiology students learn from a cure investigating the effects of septicemia on cardiac function: frog and larval *Drosophila* models. Article 69. Volume 42. In: Advances in Biology Laboratory Education. Proceedings of the 41st Conference of the Association for Biology Laboratory Education (ABLE).

Sharp, K. A., Cooper, R. L., and Carter, D. (2022). Semester-long Projects. Article 50; Volume 42. In: Advances in Biology Laboratory Education. Proceedings of the 41st Conference of the Association for Biology Laboratory Education (ABLE).

Then a number of primary research publications followed as many interesting phenomena were uncovered during the class projects. The publications from this adventure are listed below:

Ballinger-Boone, C., Anyagaligbo, O., Bernard, J., Bierbower, S. M., Dupont-Versteegden, E. E., Ghoweri, A., Greenhalgh, A., Harrison, D., Istas, O., McNabb, M., Saelinger, C., Stanback, A., Stanback, M., Thibault, O., and Cooper, R. L. (2020) The effects of bacterial endotoxin (LPS) on cardiac and synaptic function in various animal models: Larval *Drosophila*, crayfish, crab, and rodent. International Journal of Zoological Research

doi:
10.3923/ijzr.2020.33.62https://scialert.net/abstract/?doi=ijzr.2020.33.62.

These past studies led to all these:

Cooper, R. L., McNabb, M. and Nadolski, J. (2019) The effects of a bacterial endotoxin LPS on synaptic transmission at the neuromuscular junction. Heliyon-Elsevier, 5 (2019) e01430.
https://www.heliyon.com/article/e01430.

Anyagaligbo, O., Bernard, J., Greenhalgh, A. and Cooper, R. L. (2019) The effects of bacterial endotoxin (LPS) on cardiac function in a medicinal blow fly (*Phaenicia sericata*) and a fruit fly (*Drosophila melanogaster*). Comparative Biochemistry and Physiology—Part C 217:15-24.
https://www.sciencedirect.com/science/article/pii/S1532045618302424.

McNabb, M. C., Saelinger, C. M., Danley, M. and Cooper, R. L. (2019). The effects of bacterial endotoxin (LPS) on synaptic transmission at neuromuscular junction in an amphibian. International Journal of Zoological Research 15(2):38-42. https://scialert.net/current.php?issn=1811-9778.

Istas, O., Greenhalgh, A. and Cooper, R. L. (2019) The effects of a bacterial endotoxin on behavior and sensory-CNS-motor circuits in *Drosophila melanogaster*. INSECTS https://www.mdpi.com/2075-4450/10/4/115.

Saelinger, C. M., McNabb, M. C., McNair, R., Bierbower, S. and Cooper, R. L. (2019) Effects of bacterial endotoxin (LPS) on the cardiac function, neuromuscular transmission and sensory-CNS-motor nerve circuit: A crustacean model Comparative Biochemistry and Physiology, Part A. 237:110557.
https://www.sciencedirect.com/science/article/pii/S10 95643319303216.
https://doi.org/10.1016/j.cbpa.2019.110557.

Ballinger-Boone, C., Anyagaligbo, O., Bernard, J., Bierbower, S. M., Dupont-Versteegden, E. E., Ghoweri, A., Greenhalgh, A., Harrison, D., Istas, O., McNabb, M., Saelinger, C., Stanback, A., Stanback, M., Thibault, O., and Cooper, R. L. (2020) The effects of bacterial endotoxin (LPS) on cardiac and synaptic function in various animal models: Larval *Drosophila*, crayfish, crab, and rodent. International Journal of Zoological Research 16: 33-62.
doi:
10.3923/ijzr.2020.33.62https://scialert.net/abstract/?do i=ijzr.2020.33.62.

McCubbin, S., Jeoung, A., Waterbury, C., and Cooper, R.L. (2020) Pharmacological profiling of stretch activated channels in proprioceptive neuron. Comparative Biochemistry and Physiology Part C 233 (2020) 108765.
https://www.sciencedirect.com/science/article/abs/pii /S153204562030065Xdoi.org/10.1016/j.cbpc.2020.1087 65.

The past topics led students to ask questions about what type of channels are in the proprioceptive nerve ending of

the crab and crayfish models being used which produced
their manuscripts.

Istas, O., Greenhalgh, A. and Cooper, R. L. (2020).
Repetitive exposure to bacterial endotoxin LPS alters
synaptic transmission. Journal of Pharmacology and
Toxicology 15: 65-72.
doi:
10.3923/jpt.2020.65.72http://docsdrive.com/pdfs/acad
emicjournals/jpt/2020/65-72.pdf.
https://scialert.net/current.php?issn=1816-496x.

Bernard, J., Marguerite, N., Inks, M. and Cooper, R. L.
(2020). Opposing responses of bacterial endotoxin
lipopolysaccharide (LPS) and TRPA1 thermal receptors
on synaptic transmission and resting membrane potential.
Current Research in Bacteriology 13 (1):10-21.
doi: 10.3923/crb.2020.10.21.
http://docsdrive.com/pdfs/ansinet/crb/2020/10-
21.pdf.

Bernard, J., Greenhalgh, A., Istas, O., Marguerite, N. T. and
Cooper, R. L. (2020) The effect of bacterial endotoxin
LPS on serotonergic modulation of glutamatergic
synaptic transmission. Biology 2020, 9, 210. [PDF]
https://doi.org/10.3390/biology9080210.

Greenhalgh, A., Istas, O., Cooper, R. L. (2021). Bacterial
endotoxin lipopolysaccharide enhances synaptic transmission at low-output glutamatergic synapses.
Neuroscience Research 170:59-65.
doi: 10.1016/j.neures.2020.08.008.

Since bacterial infection and septicemia can result in acidosis a student wanted to address the effects of low pH on synaptic transmission and on cardiac function:

Stanley, C. and Cooper, R. L. (2021). The effect of pH on synaptic transmission at the neuromuscular junction in *Drosophila melanogaster*. Current Research in Neuroscience 11: 1-17. doi:10.3923/crn.2021.1.17. https://scialert.net/fulltext/?doi=crn.2021.1.17andorg=10.

In addressing specifically, the effects of LPS on cellular function led to these following manuscripts:

Potter, R., Meade, A., Potter, S. and Cooper, R .L. (2021) Rapid and direct action of lipopolysaccharides (LPS) on skeletal muscle of larval *Drosophila*. Biology. 2021, *10*(12), 1235; https://doi.org/10.3390/biology10121235.

Cooper, R. L. and Krall, R. M. (2022) Hyperpolarization induced by LPS, but not by chloroform, is inhibited by Doxapram, an inhibitor of two-P-domain K+ channel (K2p). International Journal of Molecular Sciences. 2022, 23(24), 15787; https://doi.org/10.3390/ijms232415787.

Ison, B. J., Abul-Khoudoud, M. O., Ahmed, S. M., Alhamdani, A. W., Ashley, C., Bidros, P. C., Bledsoe, C. O., Bolton, K .E., Capili, J. G., Henning, J. N., Moon, M., Phe, P., Stonecipher, S. B., Tanner, H. N., Taylor, I. N., Turner, L. T., Wagers, M., West, A. K and Cooper, R. L. (2022). The effect of Doxapram, a K2p channel blocker, on proprioceptive neurons: Invertebrate model. NeuroSci 2022, 3, 566-588. (2023). https://doi.org/10.3390/neurosci3040041.

Vacassenno, R. M., Haddad, C. N. and Cooper, R. L. (2023) The effects on resting membrane potential and synaptic transmission by Doxapram (blocker of K2p channels) at the *Drosophila* neuromuscular junction. Comparative Biochemistry and Physiology Part C 263 (2023) 109497. https://www.sciencedirect.com/science/article/abs/pii/S1532045622002320#ab0015.

Vacassenno, R. M., Haddad, C. N. and Cooper, R. L. (2023) Bacterial lipopolysaccharide hyperpolarizes the membrane potential and is antagonized by the K2p channel blocker doxapram. Comparative Biochemistry and Physiology Part C. 266: 2023, 109571. https://doi.org/10.1016/j.cbpc.2023.109571.

Brock, K. E., Elliott, E .R., Abul-Khoudoud, M. O. and Cooper, R. L. (2023). The effects of Gram-positive and Gram-negative bacterial toxins on cardiac function in *Drosophila melanogaster* larvae. Journal of Insect Physiology. 147 (2023):104518. https://doi.org/10.1016/j.jinsphys.2023.104518.

Hensley, N., Elliott, E. R., Abul-Khoudoud, M. O. and Cooper, R. L. (2023). Effect of 2-Aminoethoxydiphenyl borate (2-APB) on heart rate and relation with suppressed calcium activated potassium channels: Larval *Drosophila* model. Applied Biosciences 2023, 2, 236-250. https://doi.org/10.3390/applbiosci2020017.

Elliott, E. R., Taul, A. C., Abul-Khoudoud, M. O., Hensley, N. and Cooper, R. L. (2023) Effect of doxapram (a K2p channel blocker), bacterial endotoxin and pH on heart rate: Larval *Drosophila* model. Applied Biosciences 2023, 2, 406-420. https://doi.org/10.3390/applbiosci2030026.

Brock, K. E. and Cooper, R. L. (2023) The effects of doxapram blocking the response of Gram-negative bacterial toxin (LPS) at glutamatergic synapses. BIOLOGY 2023, 12(8), 1046. https://www.mdpi.com/2079-7737/12/8/1046.

Elliott, E. R., Brock, K. E., Vacassenno, R. M., Harrison, D. A., Cooper, R. L. (2024) The effects of doxapram and its potential interactions with K2P channels in experimental model preparations. Journal of Comparative Physiology A, 2024. https://doi.org/10.1007/s00359-024-01705-6.

McCubbin, S., Meade, A., Harrison, D. and Cooper, R. L. (2024). Acute lipopolysaccharide (LPS)-induced cell membrane hyperpolarization is independent of voltage gated and calcium activated potassium channels. Comparative Biochemistry and Physiology—Part C 2024,110004, ISSN 1532-0456. https://doi.org/10.1016/j.cbpc.2024.110004.

Elliott, E. R. and Cooper, R. L. (2025). Fluoxetine antagonizes the acute response of LPS: Blocks K2P channels. Comparative Biochemistry and Physiology-Part C, 287,110045. https://doi.org/10.1016/j.cbpc.2024.110045.

McCubbin, S., Abul-Khoudoud, M. O., and Cooper, R. L. (2025)The direct effects of various bacterial toxins (LPS and LTA) on membrane potential and glutamatergic transmission in a *Drosophila* model. (In review)

Hadjisavva, M. E. and Cooper, R. L. (2025) The biphasic effect of lipopolysaccharide on membrane potential. Membranes 2025, *15*, 74.
https://doi.org/10.3390/membranes15030074.

Kim, Y., Kim, J., Spedding, V. and Cooper, R. L. (2025) The impact of pharmacological and immunological interactions on light-activated ion channels and pumps. microPublication Biology.
10.17912/micropub.biology.001463

Chapter 10:

Fun with outreach projects and listening to middle school and high school students developed into research projects

Theme:
Plants

Level:
Middle School to High School

While visiting schools and giving demonstrations on electrical recording in plants with the Backyard Brains kit, students often asked if plants could communicate with each other by electrical signals. The answer provided has always been I don't know, but there are reports that chemical communication is possible by airborne factors like from the odor in the smell of cut grass. Then one day an engineering student came to our research lab and asked if we could record electrical signals between plants communicating through the soil. This is when one realizes there was a real interest in this topic not just by students in high school but graduating with engineering degrees.

So, what else was there to do but develop a lab protocol for teaching and at the same time get some primary research conducted from the efforts.

A series of papers were published and more in the works. So far this is what has come out of the adventure:

Cooper, R. L., Thomas, M., Vascassenno, R. M., Brock, K. E., McLetchie, D. N. (2022) Measuring electrical responses during acute exposure of compounds to roots and rhizoids of plants by using a flow-through system. *Methods and Protocols. 5*(4): 62. https://www.mdpi.com/2409-9279/5/4/62/htm.

Cooper, R. L., Thomas, M., McLetchie, D. N. (2022) Impedance measures for detecting electrical responses during acute injury and exposure of compounds to roots. *Methods and Protocols.* 5(4):56. https://doi.org/10.3390/mps5040056.

Thomas, M., Cooper, R. L. (2022) Building bridges: Mycelium–mediated plant–plant electrophysiological communication. Plant Signaling and Behavior 17:1. https://doi.org/10.1080/15592324.2022.2129291.

The laboratory protocol we later developed for a college level which could also be modified for high school level depending the on equipment available.

Bioelectricity in plants:
Laboratory Protocol

Designed by: Kaitlyn Brock, Matthew Thomas, D. Nicholas McLetchie and Robin L. Cooper.

Movie of the lab https://youtu.be/sEdBDbmVQ_s.

Introduction:

Ever since the discovery of bioelectricity in animals by Volta and Galvani (McComas, 2011), there has been a fascination of the origin and purpose of electrical signals in organisms. Organisms can relay information in various ways within themselves as a means of controlling cellular function and communicating among cells. This communication is commonly associated with electrical signals. These signals are associated with a change in the membrane potential of a cell which is a charge difference across a cell membrane. For example, changes in membrane potential of muscle or a nerve leads to the conduction of the electrical signal to regulate muscle contraction or relaying information to a target. Membrane potential changes occur due to the movement of ions across a membrane through protein channels. The flow of ions (or current) can be monitored by changes in the electrical field produced as a difference in electrical potential.

A common approach to measure of ionic movement across membranes of cells in skeletal muscle, heart muscle or brain using an electromyogram (EMG), electrocardiogram (ECG), and electrocardiogram (ECG). In such cases, the electrical signals are monitored over the body due to ionic movements within the associated tissue, but the field potential is detected away from the associated tissue. Not only can field potentials be detected in animals but also in plants. Movements of fluids containing ions within xylem, phloem or across cell walls also generate static and dynamic electric fields (Fromm and Lautner, 2007).

Streaming of fluid movements within compartments (i.e., tissues, cellular compartments) of a plant produces what are termed as streaming potentials (Gilbert et al., 2006; Gensler and Yan, 1988; Koppan et al., 1988; Labady et al., 2002).

Membrane potential changes within a plant can produce electrical fields which can be monitored around areal parts of the plant in the air (Pietak, 2011; Frohlich, 1968a,b, 1975) as well as around the roots (i.e., water, soil) (Love et al., 2008).

Nevertheless, measurements on the surface or within a plant provide a better ability to measure membrane potential changes as the signals are not as dissipated in the surrounding environment. Such focus measurements to assess ionic movements within the plant can potentially detect cellular activity in response to photosynthesis, injury, and response to environmental changes (ref).

The goals of this laboratory exercise is to understand how organisms can generate electrical potentials and in particular to measure electrical signals by plants and animals by various approaches to learn how such measures are made and related concepts related to bioelectricity.

Two different electrophysiological approaches are to be used and compared in this first part of the exercise in measuring electrical potentials of plants. One approach uses a standard intracellular glass microelectrode technique as a differential recording to a ground lead and another is an impedance measure to detect a change in resistance between two points which results in voltage changes when current is kept constant per Ohm's Law. The standard differential electrical measure, commonly used for animal cells, is to detect a voltage change between recording lead (glass microelectrode) and a ground lead; however, if the recording lead does not have high enough resistance, then small changes in ionic flow (current) are hard to detect. If both the ground and recording leads are immersed within a solution with a large surface exposure on the recording lead, then this will result in a low resistance input.

Thus, a reason to have a small area of contact with the tissue being measured is to have a high resistance allowing small changes in current to result in a larger voltage change as established by Ohm's law. Essentially, the smaller the area of contact the greater the resistance. A high resistance recording lead can be obtained by coating the lead with an insulation while leaving a small amount of wire exposed at the tip, or placing the recording lead with more surface exposure within a glass microcapillary which has a small tip opening to the media being measured. In this procedure, the microcapillary is filled with a conductive media such as 3M potassium chloride (KCl) or potassium acetate as typically used for recordings across cell membranes in animal tissue. For recordings within compartments (i.e., chambers such as xylem and phloem within a plant) and within cells of plants a common practice is to use 0.3 or 0.1 M KCl within the recording glass microcapillary (Yan et al., 2009; Fromm and Lautner, 2007). This first technique measures voltage where the plant produces a current and has a given resistance. Ohm's Law : $V = IR$: where $V=$ voltage, $I = $ current, and $R = $ resistance.

This first approach for electrophysiological recording is susceptible to field potentials in the environment such as 50 (Europe) or 60 (North America) Hz frequency from electronic equipment. Thus, a Faraday cage is commonly used for such recordings to shield the environmental electrical noise.

The second electrophysiological approach is an impedance measure and is similar to the differential recording mentioned above. Two leads are used to detect a change in voltage difference due to a change in resistance. This is often referred to as a measure of dynamic resistance. With the impedance measure two leads are used to detect a change in resistance while passing a constant current. Impedance measures are used in various ways such as respiratory breathing rates with expansion and relaxation of a chest for mammals (Bachmann et al., 2018), the movements of a respiratory organ in crayfish to control aeration of gills (Schapker et al., 2002), clinical neuro-muscular disease research in mammals (Nagy et al., 2019), the heart rate of crustaceans submerged in water (Listerman et al., 2000; Li et al., 2000), as well as to detect when the environment causes physiological stress of crayfish, crab or shrimp (Weineck et al., 2018).

Even the fine movements of a beating heart in larval *Drosophila* can be detected as there is a wide range in the sensitivity with an impedance technique without detecting surrounding electrical noise. Depending on how the measures are made, they can be noninvasive, such as a strap around the chest of a mammal or two leads in the media to detect body movements of insect larvae (Cooper and Cooper, 2004; de Castro and Cooper, 2020).

With two leads in a media or solution with an organism or a tissue present, a small electrical field can be used. If there is any change in the resistance between the two leads, such as the movement of ions, this will be detected.

Herein, we used these electrophysiological approaches to measure electrical changes due to ionic movement within a plant during injury and one can even exposure various compounds to the roots of the plant if time allows. Other measures can be recorded such as the response of a healthy plant to stimuli and disease states, and ionic movements within the plant that occur during metabolic processes such as photosynthesis. Such measures are not only possible for acute changes within milliseconds but monitoring long-term recordings over days, weeks, and months are feasible.

Methods of electrophysiology

List of materials needed for 1 set up:

Scissors (1)

Forceps (1)

Silver Wire for ground wire (1)

Conductive paint

Microscope (1)

Electrode Probe for intracellular measures(1)

Glass electrode and KCl solution to fill electrode

Amplifier/Acquisition System (1)

Faraday Cage (1)

Desktop/Laptop (1)

Plant anchored as not to move (1)

A micromanipulator to hold electrode probe

Electrophysiology uses a standard intracellular technique.

Measuring electrical responses within the stems of plants will be performed by inserting a glass microelectrode (catalogue # 30-31-0 from FHC, Brunswick, ME, 04011, USA) with tips broken to jagged openings in the range of 10 to 20 μM diameter. The electrode will be filled with 0.3 mM KCl). A ground wire will be placed in the moist soil next to the plant being recorded or attached to the plant with conductive paint. The electrical signals will be obtained with an amplifier (Neuroprobe amplifier, A-M systems from ADInstruments, Colorado Springs, CO. 80906 USA) and connected to a computer via an AD converter (4s Power lab 4/26, ADInstruments, Colorado Springs, CO. 80906 USA), (Figure 7). Recordings will be performed at an acquisition rate of 20 kHz. Events will be observed and analyzed with software Lab-Chart 8.0 (ADInstruments, USA).

The silver wires of the recording and ground wire will be coated with chloride by using bleach for about 20 minutes to obtain the Ag-Cl coating. All wires are rinsed thoroughly with water prior to being used. A glass electrode is to be placed within the stems with a micromanipulator under a dissecting microscope. The electrodes were inserted 1 to 2 mm into the stem of the plants (Figure 1 and 2). The recording set up is performed within a grounded Faraday cage as shown in the YouTube link: https://youtu.be/Jv7XYhu-kCs.

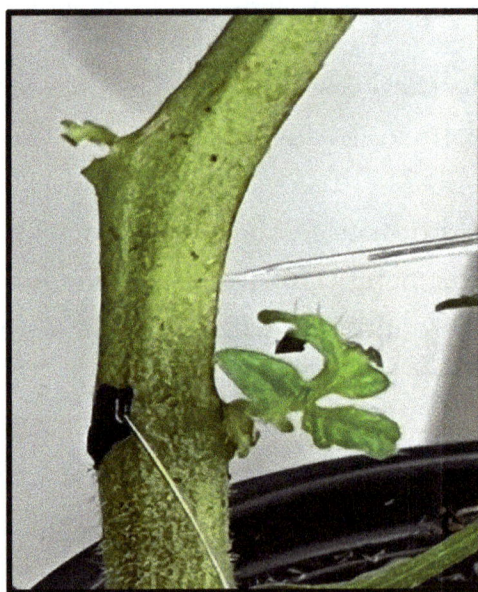

Figure 1:
A representative tomato plant
with a glass electrode placed within the stem and the
ground wire attached to the stem with conductive black paint.

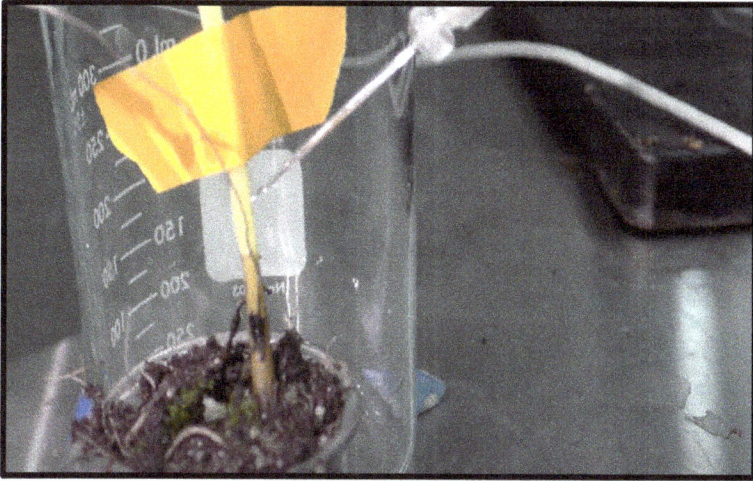

Figure 2:
**A representative *Coleus scutellarioides* with a glass electrode
placed within the stem and the ground wire attached
to the stem with conductive paint. Here the stem is taped
to a beaker for stabilizing the stem.**

Electrophysiology
using the impedance technique:

The impedance technique is used for the plant model of
choice (*Coleus scutellarioides*). Two insulated iridium-
platinum wires (diameter 0.127 mm and with the coating
0.2032 mm; A-M Systems, Carlsburg, WA, USA) or
insulated stainless steel wires (0.127 mm diameter and with
coating 0.2032 mm diameter; A-M Systems, Carlsburg,
WA, USA) can be used. The iridium-platinum wires are
more flexible than the stainless-steel wires, but the
stainless-steel wires are preferable due to the stiffness of
steel in penetrating the stem of the plant. In addition, the
stainless-steel wires are about a third of the cost. The

insulation (~ 0.5 mm length) was removed with fire on the ends of both wires to be in contact with the plant. The other ends had the insulation removed (~ 1 cm) to be placed in the clamps of the impedance amplifier. The impedance amplifier (model 2991, UFI, Morro Bay, CA, USA, Figure 10) was used, which allowed changes in an electrical field to be monitored as a measure of dynamic resistance.

Two approaches will be used for impedance measures. One approach involves placing the two leads along the stem of the plants with physical contact but not penetrating the tissue. In this case, the conductive paint was applied sparingly over the exposed ends of the wires and on the plant. A second approach was to impale the stem of the plants with both leads to a depth of about 1 mm or less. The two leads were 5 to 10 cm apart.

The output of the impedance amplifier was connected to a computer via an AD converter (4s Power lab 4/26, ADInstruments, Colorado Springs, CO. 80906 USA). Recordings were performed at an acquisition rate of 20 kHz for acute measures and at 100points/sec for long term recordings over hours. Events were observed and analyzed with software Lab-Chart 8.0 (ADInstruments, USA). Figure 3 illustrates the impaling approach with the impedance wires.

Figure 3:
Coleus scutellarioides showing the two leads
for impedance measures. The two leads are placed
about 1 mm within the stem and are about 6 cm apart.

Figure 4:
Impedance wires placed
in (penetrated) the stem of *Coleus scutellarioides.*

Stimuli

1) Injury induction

Can changes in electrophysiology as a result of mechanical injury? We will cut the leaf with some scissors to test for electrophysiological changes. Because cutting a leaf with a scissors requires the leave to move, we will test if leaf movement without cutting results in electrophysiological changes. This will be tested by bending the leaf to the same degree as would occur by cutting the leaf. Leaves are to be taped to a supporting structure to avoid any movement of the stem where the recording leads will be placed. (Figure 5 to 8).

The associated videos and figures illustrate some of the leaf bends and cuts to be performed.

https://youtu.be/sEdBDbmVQ_s.

Figure 5:
Bending a leaf while recording using impedance technique.

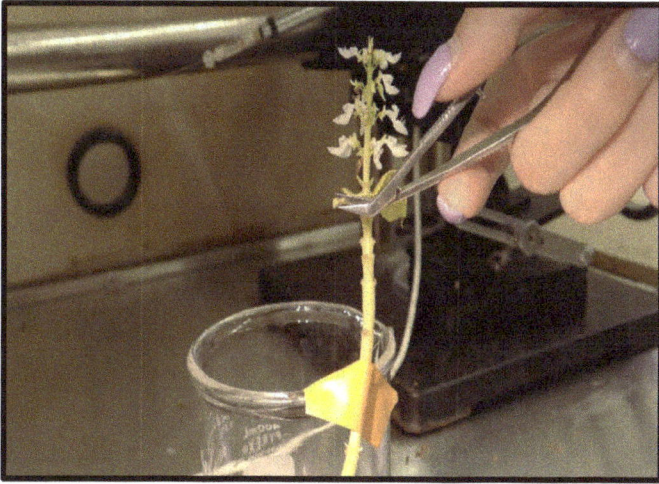

Figure 6:
Cutting a leaf while recording
with impedance technique.

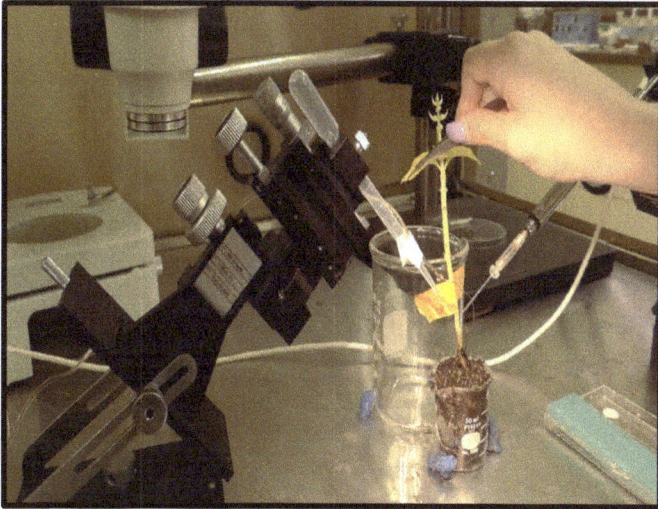

Figure 7:
Bending leaf
using glass electrode technique.

Figure 8:
Cutting leaf using glass electrode technique.

Electrical recording:

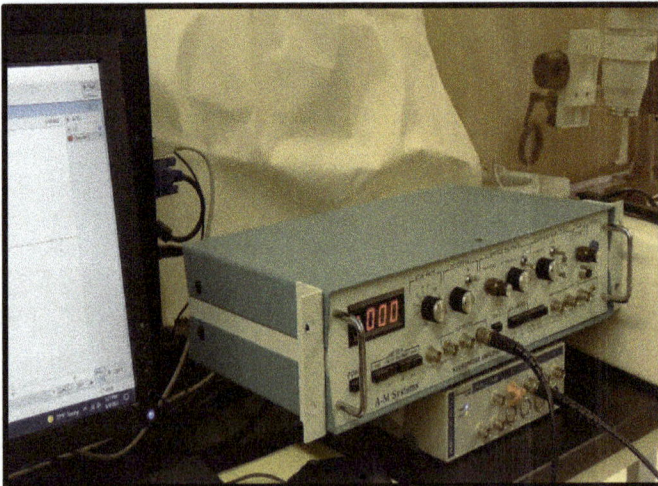

Figure 9:
The amplifier used for the glass electrode technique.

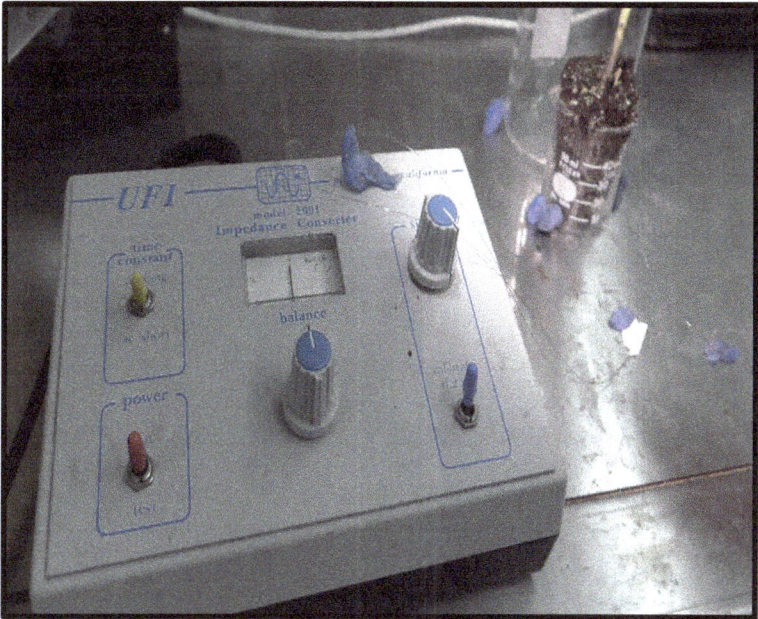

Figure 10:
An impedance converter.

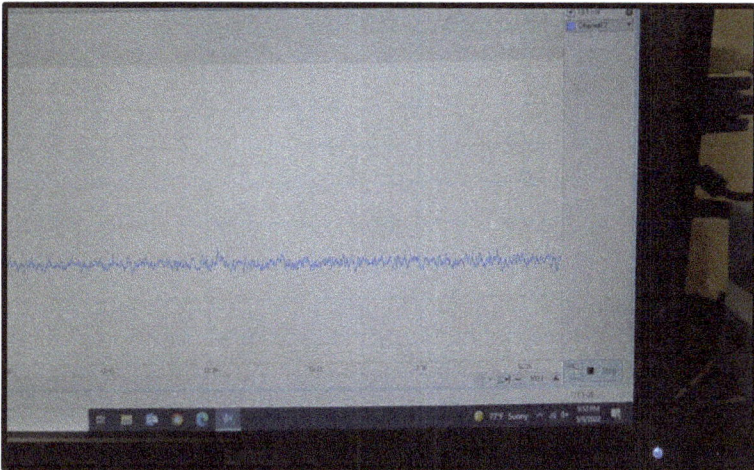

Figure 11:
Screen response from bending with impedance.
No deflection in the trace is correlated
with bending of the leaf.

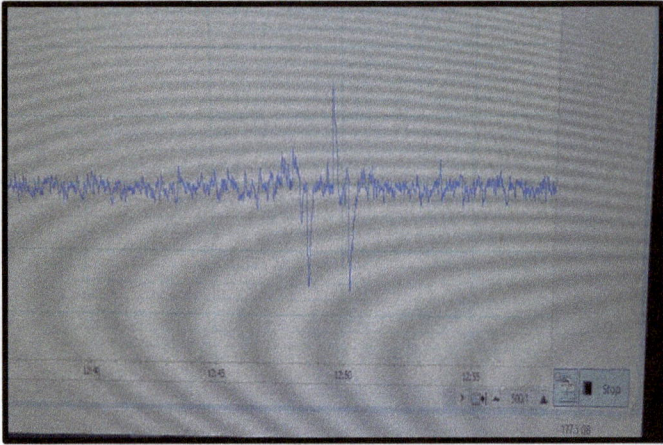

Figure 12:
Screen response from cutting a leaf
while measuring the responses
with the impedance technique.

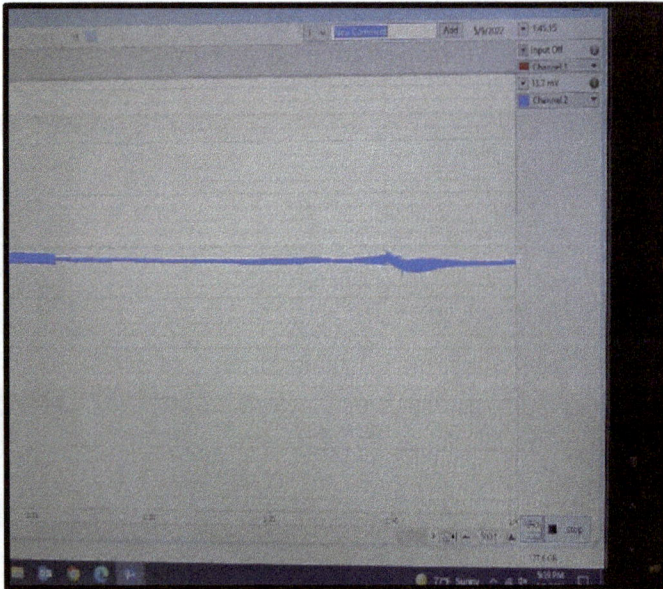

Figure 13:
Screen response from bending a leaf
while using the glass electrode.

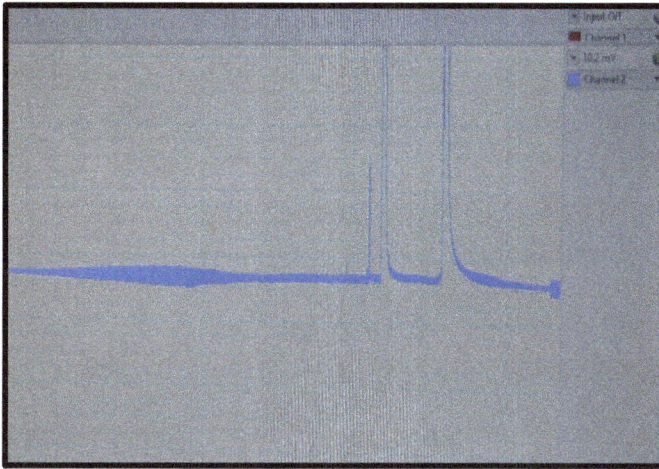

Figure 14:
**Screen response while making a cut
on a leaf while using the glass electrode.**

Analysis:

Measure the amplitude of responses related to bending a leaf and cutting a leaf with the different recording techniques. Place a marker at the base of response and move the cursor over the trace to measure the difference. Record the values in notebook and on the whiteboard for other groups and you your data.

Discussion points

1. Was there a difference between touching the leaves and cutting the leaves across both approaches?

2. Of the two approaches to measure electrophysiological response which of the two are better? Why?

3. What is causing the electrical responses in plants?

4. Is this electrophysiological information useful for the plant? If so, how do you know? How can you prove that the information is used by the plant?

5. Would different plants give different responses to same stimuli?

6. Hypothesize how plants might electrically communicicate with each other?

Part 2:
Electrical responses from animal tissue

Extracellular recording of animal tissue:

See the associated movie with this module in how to record an EMG in a human with the EMG Spiker box from Backyard Brains. https://youtu.be/sEdBDbmVQ_s.

Intracellular recording of animal tissue:

The next exercise is to record the resting membrane potential in animal tissue. Please see this detailed protocol for recording membrane potentials in crayfish muscle.

Baierlein, B., Thurow, A. L., Atwood, H. L., Cooper, R. L. Membrane Potentials, Synaptic Responses, Neuronal Circuitry, Neuromodulation and Muscle Histology Using the Crayfish: Student Laboratory Exercises. *J. Vis. Exp.* (47), e2322, doi:10.3791/2322 (2011). https://app.jove.com/v/2322/membrane-potentials-synaptic-responses-neuronal-circuitry.

References:

Bachmann, M. C.; Morais, C.; Bugedo, G.; Bruhn, A.; Morales, A.; Borges, J. B.; Costa, E.; Retamal, J. Electrical impedance tomography in acute respiratory distress syndrome. Crit. Care 2018, 22, 263.

Cooper, A. S.; Cooper, R. L. Monitoring activity of *Drosophila* larvae: Impedance and video microscopy measures. *Drosophila* Infor. Service 2004, 87,v85-87.

de Castro, N. S.; Cooper, R. L. Useful techniques for measures with *Drosophila*: larval movements, heart rate, and imaging. Methods and Protocols 2020, 3(1), 12; https://doi.org/10.3390/mps3010012.

Frohlich, H. Bose. Condensation of strongly excited longitudinal electric modes. Physics Letters A 1968, 26A, 402-403.

Frohlich, H. Long-range coherence and energy storage in biological systems. International J. of Quantum Chemistry 1968, 2, 641-649.

Frohlich, H. Evidence for bose condensation-like excitation of coherent modes in biological systems. Physics Letters A. 1975, 51A, 21-22.

Fromm, J.; Lautner, S. Electrical signals and their physiological significance in plants. Plant Cell Environ. 2007, 30(3), 249–257.
https://doi.org/10.1111/j.1365- 3040.2006.01614.x.

Gensler, W.; Yan, T. Investigation of the Causative Reactant of the Apoplast Electropotentials of Plants. J. Electrochemical Society: Electrochemical Science and Technology. 1988, 135, 2991–2995.

Gilbert, D.; Mouel, J. L. L.; Lambs, L.; Nicollin, F.; Perrier, F. Sap flow and daily electrical potential variations in a tree trunk. Plant Science. 2006, 171, 572–584.

Koppan, A.; Szarka, L.; Wesztergom, V. Conclusions from multichannel electrical recordings in a standing tree. Ann. Geophys. 1998, 16, 271.

Labady, A. J.; Thomas, D. J.; Shvetsova, T.; Volkov, A. G. Plant bioelectrochemistry: effects of CCCP on electrical signaling in soybean. Bioelectrochemistry. 2002, 57, 47–53.

Li, H.; Listerman, L.; Doshi, D.; Cooper, R. L. Use of heart rate to measure intrinsic state of blind cave crayfish during social interactions. Comp. Biochem. Physiol. A 2000, 127, 55-70.

Listerman, L.; Deskins, J.; Bradacs, H.; Cooper, R. L. Measures of heart rate during social interactions in crayfish and effects of 5-HT. Comp. Biochem Physiol A. 2000, 125, 251-264.

Love, C. J.; Zhang, S.; Mershin, A. Source of sustained voltage difference between the xylem of a potted Ficus benjamina tree and its soil. PloS One, 2008, 3(8), e2963. https://doi.org/10.1371/journal.pone.0002963.

McComas, A. J. (ed.). Galvani's Spark: The story of the nerve impulse. New York: Oxford University Press. 2011.

Nagy, J. A.; DiDonato, C. J.; Rutkove, S. B.; Sanchez, B. Permittivity of ex vivo healthy and diseased murine skeletal muscle from 10 kHz to 1 MHz. Sci. Data 2019, 6, 37.

Pietak A. M. Endogenous electromagnetic fields in plant leaves: a new hypothesis for vascular pattern formation. Electromagn Biol Med. 2011, 30(2), 93-107. doi: 10.3109/15368378.2011.566779. PMID: 21591894.

Schapker, H.; Breithaupt, T.; Shuranova, Z.; Burmistrov, Y.; Cooper, R. L. Heart rate and ventilatory correlative measures in crayfish during environmental disturbances and social interactions. Comp. Biochem. Physiol. A 2002, 131, 397-407.

Weineck, K.; Ray, A. J.; Fleckenstein, L.; Medley, M.; Dzubuk, N.; Piana, E.; Cooper, R. L. Physiological changes as a measure of crustacean welfare under different standardized stunning techniques: Cooling and electroshock. ANIMALS 2018, 8(9), 158; https://doi.org/10.3390/ani8090158.

Yan, X.; Wang, Z.; Huang, L.; Wang, C.; Hou, R.; Xu, Z.; Qiao, X. Research progress on electrical signals in higher plants. Prog. Nat. Sci. 2009, 19(5), 531–541. https://doi. org/10.1016/j.pnsc.2008.08.009.

Chapter 11:

Getting one's experiences out to educators

Examples

Educational modules are traditionally published in text and published in journals or books but now with the advent of easy to record media with a phone one can supplement text with movies on YouTube or host activities on web-based platforms. Some journals now allow supplemental movies to be attached to primary manuscripts and the publisher hosts the space.

There are journals now that publish details in experimentation as visual explanations with supplemental text. Our research group has made good use of this type of venue to present how to conduct experimental protocols. There is a cost to publishing and making the resources open access.

We have made this all-open access. Most of our published online works were made with video recordings with a simple camera or even a phone camera by our own research laboratory members.

Some are listed on the next page:

Cooper A. S., Cooper R. L. Historical view and physiology demonstration at the NMJ of the crayfish opener muscle. J Vis Exp. 2009 Nov 9;(33):1595. doi: 10.3791/1595.

Bierbower S. M., Cooper R. L. Measures of heart and ventilatory rates in freely moving crayfish. J Vis Exp. 2009 Oct 15;(32):1594. doi: 10.3791/1594.

Cooper A. S., Rymond K. E., Ward M. A., Bocook E. L., Cooper R. L. Monitoring heart function in larval *Drosophila melanogaster* for physiological studies. J Vis Exp. 2009 Nov 16;(33):1596. doi: 10.3791/1596.

Leksrisawat B., Cooper A. S., Gilberts A. B., Cooper R. L. Muscle receptor organs in the crayfish abdomen: a student laboratory exercise in proprioception. J Vis Exp. 2010 Nov 18;(45):2323. doi: 10.3791/2323.

Wu W. H., Cooper R. L. Physiological recordings of high and low output NMJs on the crayfish leg extensor muscle. J Vis Exp. 2010 Nov 17;(45):2319. doi: 10.3791/2319.

Robinson M. M., Martin J. M., Atwood H. L., Cooper R. L. Modeling biological membranes with circuit boards and measuring electrical signals in axons: student laboratory exercises. J Vis Exp. 2011 Jan 18;(47):2325. doi: 10.3791/2325.

Cooper A. S., Leksrisawat B., Gilberts A. B., Mercier A. J., Cooper R. L. Physiological experimentation with the crayfish hindgut: a student laboratory exercise. J Vis Exp. 2011 Jan 18;(47):2324. doi: 10.3791/2324.

Wu W. H. , Cooper R. L. Physiological recordings of high and low output NMJs on the crayfish leg extensor muscle. J Vis Exp. 2010 Nov 17;(45):2319. doi: 10.3791/2319.

Baierlein B., Thurow A. L., Atwood H. L., Cooper R. L. Membrane potentials, synaptic responses, neuronal circuitry, neuromodulation and muscle histology using the crayfish: student laboratory exercises. J Vis Exp. 2011 Jan 18;(47):2322. doi: 10.3791/2322.

Titlow, J., Majeed, Z. R., Nicholls, J. G. and Cooper, R. L. (2013). Identifiable neurons in the central nervous system of a leech via electrophysiology and morphology, sensory field maps in skin and synapse formation in culture: Student laboratory exercises. Journal of Visualized Experiments (JoVE). (81), e50631, doi:10.3791/50631.

Titlow, J., Majeed, Z. R., Hartman, H. B., Burns, E., and Cooper, R. L. (2013). Neural Circuit Recording from an Intact Cockroach Nervous System. Journal of Visualized Experiments (JoVE). (80), e51050, doi: 10.3791/51050.

Majeed, Z. R., Titlow, J., Hartman, H. B. and Cooper, R. L. (2013). Proprioception and tension receptors in crab limbs: Student laboratory exercises. Journal of Visualized Experiments (JoVE). (80), e51050, DOI: 10.3791/51050.

Epilogue

We have found it easier to publish homemade (i.e., lab made) movies at no cost and host them on the university server under a faculty member's account or a course web page or within a course web-based program, such as blackboard or canvas. The downside of using a university class-based program is that it is limited to only the class.

Also, hosting on a university server under a faculty member's account is problematic since when they retire or leave that particular institution the web site access will likely be dropped and no longer be publicly accessible. One might pay an outside source but that is only as long as one continues to pay for the service. We have also posted many class projects on YouTube as noted in the above modules.

Self-publishing a book of a project or an educational manual may also be a suitable means to allow public access for a long period of time. In some cases, this might even be less expensive and speeder than submitting to a peer reviewed journal. A past experience which led us to publish this book was that we submitted a manuscript to be published in a journal which was aimed at promoting the use of an alternative model animal for teaching hands-on experimental laboratory procedures. The protocol was first submitted to the media to promote undergraduate scholarly work within the university. The selection committee was not interested in an educational protocol for one of the university's own courses.

So, then the manuscript was submitted to a journal which specializes in protocols, but the reviewers' wanted results published on the use of the protocol and student assessment. The protocol was not meant to be a research paper, and it was not meant to be a long drawn-out procedure to obtain IRB approval of student assessment to conduct the protocol. In addition, the publisher of that journal wanted ~$2,000 USD for making the content open source. Thus, we sought out a publisher to help us self-publish and it was published in two weeks on Amazon books and made accessible for $1.99 (USD) to download an electronic copy from Lulu.com website. Any proceeds go to the student author.

Kaitlyn E. Brock and Robin L. Cooper (2024). The Lobster: A Model for Teaching Neurophysiological Concepts. Azalea Art Press. Sonoma, California, USA.

Available through Lulu.com as an E-book and Amazon as well as Barnes&Noble. ISBN: 978-1-943471-84-3.

https://www.lulu.com/shop/robin-l-cooper-and-kaitlyn-e-brock/the-lobster-a-model-for-teaching-neurophysiological-concepts/paperback/product-zm8gyq2.html?q=The+Lobsterandpage=1&pageSize=4

For print book and E-book orders please visit: www.Lulu.com.

The hope of presenting this jumble of educational activities of various sorts serves a number of purposes. One is that maybe others will find some use for the activities. Secondly, the goal was to bring attention to some variety of educational activities and where to find many of them that are vetted and tested such as in publications by ABLE or other primary papers. Thirdly, to prompt being open minded about conducting outreach STEM activities in elementary, middle and high schools as well as public venues as such activities to promote science education to the public. Ideas may come to mind by the simplest questions people ask which can turn into productive research projects. Fourthly, teaching college CURE classes at a freshman level as well as advanced level is very rewarding not only for the students but also for the instructor. If the instructor tries, the CURE projects can be turned into ACURE projects and may even result in primary publications. Such projects can also stimulate students to continue investigating a project they started.

We have had students start projects from high school activities in the laboratory and end up staying in the lab all through 4 years of undergraduate training. We have even had undergraduate students who went on to medical school to come back to the lab to conduct a research elective during their 4th year of medical school. And lastly one can appreciate the flow of educational projects into research as the process in the nature of science and scientific discoveries. By being an educator at various primary and secondary levels one can also be productive in research.

— *Robin L. Cooper*
September, 2025

About the Author

Dr. Robin L. Cooper obtained a dual B.S. in Chemistry and Zoology from Texas Tech University in 1983 and a Ph.D. in 1989 in Physiology from the School of Medicine, and postdoctoral training (1989-1992) at the University of Basel, School of Medicine, Basel, Switzerland. His second postdoctoral stint (1992-1996) was at the Department of Physiology, University of Toronto, School of Medicine, Toronto, Canada. In 1996, he joined the Department of Biology at the University of Kentucky and is now a Professor. He also obtained a BSN in nursing in 2012 and practiced as an RN from 2011 to 2017. He has received several teaching awards and continues to mentor students in research based activities and publishing for peer review. In his spare time, he FaceTimes with his first grandchild, Rose, and cycles the backroads of Kentucky.

To Contact the Author:
Robin Cooper
RLCOOP1@UKY.edu

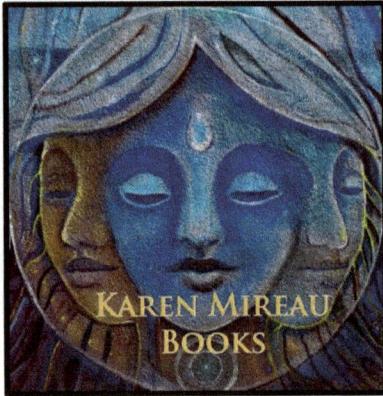

Karen Mireau
Books

To Contact the Publisher:
Karen Mireau
KarenMireauBooks@gmail.com

www.ingramcontent.com/pod-product-compliance
Lightning Source LLC
Chambersburg PA
CBHW072125090426
42739CB00012B/3065